PRACTICAL PROCEDURES
IN DIAGNOSTIC RADIOLOGY

PRACTICAL PROCEDURES
IN
DIAGNOSTIC RADIOLOGY

By

H. M. SAXTON

M.B., M.R.C.P., F.F.R., D.M.R.D.
Consultant Radiologist, Guy's Hospital, London

and

BASIL STRICKLAND

F.R.C.P., F.F.R., F.A.C.R. (Hon.)
Consultant Radiologist, Westminster Hospital, London
Consultant Radiologist, Brompton Hospital, London

SECOND EDITION

With 48 illustrations

LONDON
H. K. LEWIS & Co. Ltd.
1972

First Edition 1964
Second Edition 1972

PRINTED IN GREAT BRITAIN
FOR H. K. LEWIS & CO. LTD., 136 GOWER STREET, LONDON
BY ROBERT MACLEHOSE & CO. LTD.
THE UNIVERSITY PRESS, GLASGOW

With Contributions by and assistance from the following:

PETER ARMSTRONG, M.B., F.F.R., Consultant Radiologist, King's College Hospital, London.

ELLIS BARNETT, F.F.R., D.M.R.D., Consultant Radiologist, Western General Hospital, Glasgow.

J. L. BOLDERO, B.M., B.Ch., F.F.R., D.M.R.D., Consultant Radiologist, United Oxford Hospitals.

WILLIAM CAMPBELL, M.D.Edin., D.M.R.D., Consultant Radiologist, Queen Victoria Hospital Plastic Surgery and Jaw Injuries Centre, East Grinstead.

J. O. CRAIG, F.R.C.S.I., F.F.R., D.M.R.D., Consultant Radiologist, St. Mary's Hospital, London and Bolingbroke Hospital, London.

DAVID EDWARDS, M.B., F.R.C.P., F.F.R., D.M.R.D., Consultant Radiologist, University College Hospital, London.

STANLEY FELDMAN, M.B., B.S.(Honours), F.F.A.R.C.S., D.A., Consultant Anaesthetist, Westminster Hospital, London.

IAN ISHERWOOD, M.B., B.Ch., F.F.R., D.M.R.D., Consultant Radiologist, Manchester Royal Infirmary.

LIPMANN KESSEL, *M.B.E.(Mil.)*, *M.C.*, F.R.C.S., Consultant Orthopaedic Surgeon, Fulham, St. Mary Abbots & Western Hospitals, London.

L. KREEL, M.D., M.R.C.P., F.F.R., Head of Radiology, Northwick Park Hospital and Clinical Research Centre, Middlesex.

B. W. LACEY, M.D., B.Sc., Professor of Bacteriology, Westminster Hospital Medical School, London.

W. F. WHITE, M.B., B.S., F.F.R., Consultant Radiotherapist, St. Luke's Hospital, Guildford.

C. G. WHITESIDE, B.M., B.Ch., F.F.R., D.M.R., Consultant Radiologist, Middlesex Hospital & Royal National Nose, Throat and Ear Hospital.

F. W. WRIGHT, B.M., B.Ch., M.R.C.P., F.F.R., Consultant Radiologist, United Oxford Hospitals.

W. B. YOUNG, *M.B.E.*, F.R.C.S., Ed., F.F.R., D.M.R.D., Director, X-ray Department, Royal Free Hospital, London.

PREFACE

THIS book is intended principally for the D.M.R. trainee, the Registrar who is beginning to enlarge his practical experience and the post-graduate from overseas who, having obtained the D.M.R., is returning to his own country with very little training in special techniques. It is based on our own personal tuition of Registrars at the Westminster Hospital and our aim has been to provide those small but important details of technique normally learnt by word of mouth. It also springs from the wish to write down for use in our own department details of the less familiar examinations and to this extent we hope it may be useful on occasion to more senior radiologists. Although we have drawn mainly on our own experience this has been materially enhanced by discussions with colleagues too numerous to acknowledge individually but to whom we are most grateful. In general we have tried to select those investigations which might be expected to be performed in any well-equipped general department but it has not always been easy to decide which procedures to include and our choice has in places been somewhat arbitrary. We have left out certain examinations because they are seldom used (e.g. contrast mammography, seminal vesiculography) but we have included others which are little known in the belief that their potentialities have not yet been fully explored. To discuss the use of barium adequately would require a book in itself and this we have not attempted. Neuroradiology and the more advanced forms of vascular radiology have also been omitted, in part for reasons of space and in part because they are best learned under supervision and practised in specialised units. We have however, felt justified in including the simpler forms of vascular investigation since they are becoming standard procedures which might be requested of any radiologist with a training in present day methods. We do not in fact consider it wise for anyone to begin to carry out arteriography without some supervised training and our accounts in this section are intended to be supplementary to such tuition and to fill in any gaps which it may have left.

Wherever we thought it helpful we have given references to further reading but we have not sought to cover the literature at all extensively. To discuss all methods for any one investigation is impracticable and would detract from the value of a handbook. In the main, therefore, we have presented techniques which we personally have found satisfactory. We realise that all such techniques are continually changing but we believe that it is better to give the fullest possible descriptions of one particular method of carrying out an investigation (which can then be modified as required) than to discuss numerous techniques without giving a really comprehensive picture of any one of them.

The fascination of radiology lies in the combination of intellectual and practical interest which it can offer. Our hope in writing this book has been to help radiologists to extend the range of procedures in which they are competent. If they can do so they will add to their own satisfaction in their job and increase their value to patients and colleagues.

H. M. S.
B. S.

December 1963

PREFACE TO THE SECOND EDITION

THE purpose of this book has not altered but changes in established methods and the advent of fresh techniques have necessitated many alterations in the content. We wish to thank all our colleagues who have contributed, directly or indirectly, to this edition; in particular we thank Dr. Peter Armstrong and Sister F. I. Cole who have read the manuscript and offered helpful criticism.

H. M. S.
B. S.

June 1971

EXPLANATORY NOTES

1. Contra-indications.—It is not practicable to mention every contra-indication to an examination but most of those which we have not included are self-evident—e.g. 'toxaemia' from infection or sepsis in the area in which a needle is to be inserted. Even those we have listed are not all absolute contra-indications. In many cases they are intended to remind the radiologist that a given condition introduces a risk into the examination so that he may consider whether the likely diagnostic 'return' is worth the risk involved.

2. Local anaesthetic.—Unless otherwise stated the term 'local anaesthetic' means Lignocaine 1% for injection. We have not always mentioned the need to draw back on the plunger of the syringe before injection of local anaesthetic. This is only because such reiteration would be tedious; this practice must always be followed if local anaesthetic is to be given with safety.

3. Needles.—The terminology used for specifying needle sizes awaits rationalisation and with the publication of British Standard 3522:1962 this process has begun. The widespread adoption of disposable needles has also contributed to considerable and welcome reduction in the number of needles now available. A brief explanation of present nomenclature is given in the General Appendix together with tables showing the actual measurements of the various sizes and gauges.

4. The Chapter Appendices.—For the convenience of our readers each chapter has an Appendix which contains a variety of practical items such as preparation and aftercare of the patient, trolley settings and notes on the equipment used. Where certain instruments are not likely to be available from all surgical instrument makers we indicate, by initials, the manufacturers whom we know to supply them. This does not necessarily imply that they are the only firms to supply these articles but simply that they are the firms known to us.

Key to the initials:
A.H.—Allen and Hanbury Ltd., Bethnal Green, London, E.2.
G.U.—Genito-Urinary Mfg. Co. Ltd., 28A, 33 and 34 Devonshire Street, London, W.1.
M.D.—V. J. Millard, 36 Highgate Hill, London, N.19.
P.P.—Portland Plastics Ltd., Bassett House, Hythe, Kent.

S.X.—Sierex Ltd., 15 Clipstone Street, London, W.1.

B.D.—Becton Dickinson Ltd., York House, Empire Way, Wembley, Middlesex.

5. Tray or trolley settings.—For most procedures a tray or trolley is prepared with the instruments, etc., which are required set out upon it. Trays are adequate for the simpler investigations such as intravenous urography. When, in each chapter appendix, we give the 'setting' it is described in two parts:

'Sterile-tray' or 'Sterile-upper-shelf' are headings used to indicate the items which must be sterile and which are laid out, either on the trolley top—being later covered by a sterile towel—or in a tray. In the latter case the sterile towel is subsequently doubled over the sterile objects.

Under 'lower shelf' or 'back of tray' are enumerated items such as ampoules of drugs or contrast, which are not sterile but which are necessary for the examination. Where sterile packs are employed the arrangement of items may be different because it is then possible to keep an assortment of sterile equipment, e.g. catheters, adaptors or needles, in their own individual packs on the lower shelf. These are then available to meet the especial needs which may arise during a particular examination.

These settings are intended as a guide, particularly for those undertaking an examination for the first time or in an X-ray department to which a procedure is new. They need not necessarily be followed rigidly and it may well be found possible to use more limited settings. However, we would recommend starting with a reasonably full setting and only discarding items if experience shows that this is possible.

6. Film sizes.—When we refer to the size of a film or a cassette we give its dimensions in inches, e.g. $12'' \times 10''$. Table 9 in the General Appendix shows the exact equivalent in centimetres of the commoner film sizes. It also gives the dimensions of the corresponding film sizes used on the Continent and now coming into use in this country.

7. Sensitivity tests.—It is common when discussing procedures involving administration of iodine containing contrast media, to include 'sensitivity tests' among the suggested preliminaries. In fact, these are of no practical value, may give a misleading sense of security and are used by very few centres. The editors have therefore omitted such conventional references to sensitivity tests in all chapters. We would, however, emphasise the desirability of enquiring for a history of allergy and the vital importance of having a fully equipped resuscitation trolley and of being familiar with its use (see Chapter 3).

TABLE OF CONTENTS

SECTION III—THE BILIARY TRACT

SECTION IV—MISCELLANEOUS INVESTIGATIONS

SECTION V—VASCULAR RADIOLOGY

GENERAL APPENDIX

PRACTICAL PROCEDURES

IN

DIAGNOSTIC RADIOLOGY

CHAPTER 1

THE PATIENT IN THE X-RAY DEPARTMENT

IN this chapter we remind the reader of various points which contribute to the mental and physical comfort of the patient in the X-Ray Department and help to ensure the most effective use of X-rays in examining him. Most radiologists have overlooked such points at some time and we make no apology for stressing what is familiar to many. It is only by continued attention to every kind of detail that each examination can become as pleasant for the patient and as diagnostically rewarding as possible (see also Wild and Evans, 1968).

Waiting.[1]—Some waiting is inevitable both before and after an examination. Therefore comfortable seats or couches should be provided, in adequate warmth and light and with magazines and journals on hand. Each department will arrange a system designed to reduce waiting before an examination but the radiologist himself may be responsible for delays after the examination is completed if he does not train himself to see that the patient is sent away when the films have been passed, giving explicit instructions to the radiographer or nurse concerned.

When an examination involves further films after an interval—e.g. barium follow-through or after-fatty-meal—it is sometimes useful to tell the patient as well as the radiographer. For example one may say 'You are due to have another film taken at ten-thirty; if nobody comes for you please remind the nearest radiographer'—This helps to prevent patients being forgotten in busy departments.

The Right Patient.—A simple but valuable habit is to make a practice of asking the patient his name before starting the examination. Periodically one will find a patient who has answered to some other name than his own when called by the radiographer.

In the X-Ray Room.—The worst discomfort for many patients, particularly the thin and the elderly, is to lie on a cold, hard, bare, X-ray couch.

---

[1] For a discussion of the problems involved in reducing the waiting time see *Towards a Clearer View. The Organisation of Diagnostic X-ray Departments* (Oxford University Press, London).

1

Plastic foam mattresses reduce this discomfort and should be used whenever possible, especially for the more prolonged investigations. They must be kept in a polythene cover so that spilt contrast media can be wiped off. When a cotton cover also encloses the mattress it can be used to slide the patient over the table top during positioning.

The patient should be warm and the light should not shine into his eyes. Kyphotic patients appreciate a second pillow and some patients with backache like a supporting pad in the small of the back. In vascular disease the feet are sometimes tender and the heels are then supported on pads (or the ankle is supported in the prone position). Making the patient comfortable is important not only because it is kinder but because a comfortable patient is more likely to keep still.

All X-ray rooms should have small mobile platforms or steps so that patients can climb easily on and off the couches. When screening enfeebled patients in the erect position, hand-grips are attached to the couch so that the patient can support himself. Even a small tilt away from the vertical can make standing easier for such patients.

Explanation of the procedure.—This elementary courtesy is too often neglected. Detailed explanations are seldom necessary but most patients are reassured by being told something of what lies ahead. Before uncomfortable procedures it is better to let the patient know that there will be moments of discomfort since he does not then imagine that something has gone wrong when he feels pain. When using noisy apparatus warn the patient about the noise and of any loud calls or shouts—e.g. 'shoot' or 'take'—which will be made. More wakeful patients often like an estimate of how long the examination will last. It is best to err towards overestimating since procedures often take longer than expected and should they not do so patients will be glad to get away 'early'.

When the examination may produce unpleasant sequelae—e.g. haematoma formation in arteriography—the patient must be warned of the possibility before the examination. He will then be less likely to become either alarmed or resentful, should it occur.

The patient's consent to investigation.—Many of the more complex radiological investigations carry a certain risk and the advice of the Medical Defence Union has therefore been sought upon the need to obtain the patient's consent in writing before such examinations. A fuller discussion of the problem is given in their pamphlet 'Consent to Operative Treatment'[1] but the points bearing on radiological procedures may be summarised as follows:—

The patient's consent may be express or implied; either of these forms of consent would be equally effective in defeating an action for assault. From the legal point of view it is essential that the person consenting must understand to what he is giving consent. The nature and effect of the proposed investigation should be explained to the patient in non-technical

---

[1] Obtainable from the Medical Defence Union, Tavistock House South, Tavistock Square, London, W.C.1.

language before the investigation is begun. For simple examinations only a brief explanation is required but when the proposed examination carries a definite risk the explanation should be fuller, thus allowing the patient the opportunity of giving or withholding his consent. Ordinarily, written consent is only necessary in those cases where the investigation is carried out under a general anaesthetic but it is advisable to extend this practice to include the more hazardous procedures particularly angiographic procedures even when they are performed under local analgesia. If written consent is not obtained in such cases the procedure should be explained in the presence of a third party. In either event the explanation should be given before any sedative has been administered since the patient must be fully conscious.

Valid consent as explained above provides an effective answer against a claim for assault. Any such claim arising out of an X-ray examination is likely to be directed against the radiologist. It is therefore in the interest of every radiologist to ensure that the routine preparation for examination includes an adequate explanation at whatever stage is appropriate.

Holding the breath.—Many films are spoilt by movement due to breathing and this is frequently avoidable. In any procedure involving screening of the abdomen or chest the routine check on diaphragmatic movement can be combined with tuition in breath-holding. After the patient has been told to breathe in and then breathe out he is instructed to stop breathing and the diaphragm is watched to make sure he obeys the instructions. If he fails to hold his breath he should be made to practise.

Even when no screen control is possible it makes a great difference if the patient is given a preliminary explanation of what is to be asked of him and if one continues to exhort him during the exposure. Thus a series of encouragements is used such as 'breathe in—breathe out—stop breathing —quite still—don't breathe—don't move—just as you are'. Towards the end of this the exposure can be made in some confidence that the patient will have understood. This advice applies to the average patient but the deaf or confused may be muddled by a string of half-comprehended instructions and a single 'Hold your breath' with careful watching is the best method for such patients.

One further measure which can be of assistance to enable the patient to hold his breath is to instruct him to hyperventilate for about 20–30 seconds before the exposure. Hyperventilation is normally followed by a period of apnoea and during this period the patient can more readily hold his breath while the exposure is made.

Intravenous injections.—Skill at venepuncture is sometimes regarded as a gift but respect for certain rules will make it easier and less uncomfortable for the patient.

Finding a vein.—It is a mistake to attempt to puncture a vein which cannot be properly seen or felt. It is therefore worth expending considerable time and trouble to make certain that adequate compression has been applied so as to cause the veins to become fully engorged. If the normal manual or tourniquet compression fails a sphygmomanometer cuff should be left on the arm for five minutes at a pressure just below the diastolic; this will usually provide full venous engorgement. Sometimes it may help to place a hot-water bottle on the limb or to immerse the limb in hot water for a few minutes. Superficial veins, e.g. scalp veins or those on the dorsum of the hand or foot, are often made more prominent by tapping sharply.

Veins in fat people, well embedded in fat, are steadier to puncture than the more obvious ones in thin patients. Palpation of such veins is made easier if the skin is moistened with spirit or other sterilising fluid since the fingers slip across the skin more smoothly. When attempting to puncture mobile veins it is helpful to find a 'fork' in a vein and direct the needle at the point of junction of the two tributaries.

At the site of a recent venepuncture a vein apparently well engorged may in fact be thrombosed and should first be examined without arm compression to make certain that this is not the case.

If a vein cannot be found in one of the usual situations other possible sites are:—

The femoral vein.—The hip is extended and slightly abducted. The femoral artery is palpated. The vein lies just medial to the artery. The needle, which should be a long intra-muscular needle, is advanced cephalad at about 45° to the skin. It is thrust in to approximately the depth of the artery and then slowly withdrawn while pulling back gently on the plunger of the syringe (see also chapter on I.V. Cavography). This method is contra-indicated in infants because of the risk of producing bone or joint infection or arterial thrombosis.

The jugular and scalp veins.—Infants are held on one side with arms pinioned and the head down. This, and the accompanying crying, will engorge the scalp and neck veins. When injecting scalp veins it is necessary to shave an area of the scalp above one ear. A scalp vein set is employed.[1] The fluid to be injected is brought to the needle tip, expelling all air. The needle is held in the tip of a pair of artery forceps.[2] With the infant immobilised, the needle is gently inserted under the scalp and then introduced into the vein. An assistant draws back on the plunger of the syringe and once the position is confirmed, injects the fluid while the operator holds the needle steady. It should be pointed out that even in very small infants the antecubital veins can often, with compression, be made to stand out well and may be easy to puncture. They should always be inspected before concluding that scalp vein puncture is necessary.

In adults who have collapsed, the head-down position will commonly provide adequate engorgement of the jugular veins.

Size of Needle.—When injecting large volumes it is advisable to use a

---

[1] The usefulness of a scalp vein set is not confined to the scalp veins; in any situation where movement might dislodge a needle the scalp vein set will be found of value.

[2] The 'Butterfly' scalp-vein needle (Abbott) is made with a pair of flexible flanges by which it is easily held and manipulated; it is the best type currently available.

reasonably large needle e.g. Gillette thin wall 19G–1. R or S–TW. Eccentric nozzles on the syringes also contribute to the ease of venepuncture.

Puncturing the vein.—The time spent traversing the skin should be as short as possible to reduce pain. The needle is jabbed quickly through the skin which has been made taut by drawing it slightly towards the needle point. Commonly it is possible to enter a well-fixed vein in a single thrust.

Other points worth noting are:—

1. Wait until the needle has been attached before expelling air-bubbles from the syringe. A blocked needle will then be discovered in time.

2. Chlorpromazine (Largactil) and promethazine (Phenergan) may cause considerable pain if they are inadvertently injected into an artery. Although there is not the same risk of thrombosis as with pentothal it is advisable to leave the needle in the artery until 3 ml. of 1 % procaine or lignocaine have been injected to relieve the pain.

3. When the needle is removed from the vein press on the puncture site with a dry swab and then lift the arm to the vertical position, the hand resting against the X-ray tube. This causes the arm veins to collapse and prevents oozing and haematoma formation. Three minutes is normally sufficient.

Preparation of the bowels.—The need for a completely reliable method of clearing the small and large bowel of gas and faeces is felt by all practising radiologists. For most of the examinations discussed in this book the standard of preparation required is less exacting than that demanded for double-contrast enema; nevertheless, in many examinations, e.g. selective arteriography of abdominal vessels or in pelvic pneumography, the presence of gas and faecal shadows can cause great difficulty in obtaining definitive films. In others, such as urography, it is usually possible to obtain satisfactory films by tomography but this involves extra irradiation. The tendency to gas accumulation in the small and large bowel is most marked at the extremes of life but no useful methods are available for infants and small children and the remarks in this section apply mainly to adults and older children.

The use of oral laxatives in bowel preparation is only partly effective in most subjects but continues to be used because the alternative is some form of enema. The type of enema given ranges from a simple enema saponis to the highly effective colonic irrigation provided by the Henderson Colonic Lavage apparatus when used by an experienced nurse. The choice of method used for preparation will depend on the investigation which the patient is to undergo, the apparatus and staff available to carry out the preparation and the age and condition of the patient. For in-patients it is advisable to arrange a routine to ensure that the house officer in charge of the case is asked to review the suggested preparation before it is instituted; this may avert the use of colonic irritants in subjects with conditions such as ulcerative colitis.

Our own regimes which have proved satisfactory for most cases will be described as well as the Dulcolax or Dulcodos regime which, although less complete in its effect, is widely used and is regarded as reasonably adequate. Apart from the measures directed at the colon itself it is important to see that in-patients are kept up and about as much as possible.

In-patient preparation.—Two Dulcolax or Dulcodos tablets the night before the examination. Clysodrast enema or Veripaque 1 hour before X-ray.

Out-patient preparation.—Two Dulcolax or Dulcodos tablets the night before the examination. Enema saponis 1 hour before X-ray.

Dulcolax preparation.—Two Dulcolax or Dulcodos tablets the night before the examination. Dulcolax suppository 1 hour before the examination, *or* 2 Dulcolax tablets for 2 nights before the examination.

Apart from bowel preparation some general measures may be suggested. The first is that patients should be as active as possible so as to reduce intestinal gas. A low residue diet may have some effect. And where a patient may be apprehensive sedation is valuable in reducing air swallowing.

Note.—1. When patients are elderly and bedridden or when they are known to be constipated, it is advisable to extend the preparation over the 2 or 3 days preceding the examination, i.e. Dulcolax tabs. 2 for 2 or 3 nights before.

2. Radiologists will be able to advise nurses on the technique of enema administration since they watch the process so frequently. After the first few ounces of fluid have run in there is often a sensation of discomfort as the rectum becomes distended. The enema is stopped and the patient is strongly encouraged to hold the enema in, being told that the discomfort will pass off in a short while. When it does so instillation is resumed. Whenever the patient feels further discomfort the enema is stopped. Deep breathing and rolling into the prone position and onto the right side are valuable in persuading the enema fluid around the colon. Patient administration of small amounts will usually enable even the most constipated patient to be filled adequately, i.e. with at least 2 pints (or 1 litre).

Such advice is not needed by the experienced nurse and there is no doubt that the best results are obtained by nurses who are regularly administering enemas.

Trickle enema.—In hospitals where the X-ray department is without nursing staff to carry out enemata or where ward preparation is variable in quality, the 'trickle enema' may offer some improvement. The principle (Scott Harden, Philp & Moule, 1967) is that the enema fluid runs in slowly so that the discomfort often produced by distension of the bowel is minimised and the fluid has the best chance of reaching the caecum.

The method can be applied to patients on the ward or in the X-ray department. The enema is given from a disposable bag suspended about 2′ 6″ (80 cm.) above the bed or couch. A method of providing a 'choke' or narrow aperture in the tube is needed—one suitable 'choke' is a plastic sheath for a disposable needle with a fine hole pierced in it using a red hot 23 gauge needle. Whatever system is used it should deliver 3 pints of fluid, e.g. Veripaque or Clysodrast in 7–10 minutes. The fluid may be run in with the patient prone or supine; as it runs in he should rock from side to side and finally should lie onto the right side until unable to restrain the call to stool. Plenty of time should be allowed for evacuation and the enema should be given $1\frac{1}{4}$ hours before the examination is due.

Consultation with clinicians.—Most requests for an X-ray examination are reasonable and valid. However, from time to time it does happen that requests are made for examinations which:—

1. Are not necessary at all.
2. Have been adequately performed elsewhere.
3. Are inadvisable in the given circumstances.
4. Are not the most suitable for the particular clinical problem.

The radiologist should therefore encourage his clinical colleagues to discuss problems with him in advance. Furthermore, requests for special examinations (if not all requests) should be seen well before the examination is due to be carried out. Any problems which then arise can be settled in good time; if necessary the patient should be examined in the ward by the radiologist who should take nothing for granted but should verify the clinical signs for himself.

Reduction of irradiation.—The importance of adequate protection against irradiation for both patients and X-ray staff is well recognised. In order to save space we have made very little mention of protection in our accounts. It is assumed throughout that all possible measures are taken and that X-ray sets are checked by a competent safety officer. Protection of X-ray department personnel by wearing protective aprons, keeping as far as possible from the field of irradiation and the use of lead-containing screens will be familiar to our readers. So will the everyday measures to reduce dosage to the patient—shielding of gonads, use of light-beam delineators, high kilovoltage techniques, image intensification and so forth. But as well as all these the radiologist must cultivate the attitude of mind which sees protection as meaning *protection against unnecessary irradiation*. Viewed in this way protection covers a much wider field and includes many of the points made earlier in this section. A patient is protected against unnecessary irradiation by seeing that the examination most appropriate to his clinical problem is carried out; by making sure that he does not move or breathe during exposure, so preventing the need for repeat films; by efficient cleansing of the bowels, ensuring that an urogram is entirely adequate so that a retrograde pyelogram or repeat urogram are not needed. The use of low kilovoltages or standard films is justified by this argument when they produce better and more conclusive films; similarly the additional view which converts a suspicion into a certainty may save a great deal of further X-ray examination.

In this connection we would also emphasise the importance of the preliminary film. It may seem paradoxical to include the plain film among the measures tending to decrease irradiation but experience shows that it adds considerably to the efficiency of an examination and so to the more economical use of X-rays in diagnosis. Among the reasons for taking a plain film are:—

1. To provide a basis for the interpretation of the films taken following introduction of contrast.

2. To show abnormalities, e.g. a calycine stone, which might be concealed by the contrast later in the examination.
3. The plain-film findings may contra-indicate the examination, e.g. residual barium in the colon before an intravenous cholangiogram; or they may point to the need for special techniques, e.g. gas in the colon indicating that tomography will be of value in an intravenous urogram.
4. Without the guidance provided by the plain film there is no certainty that the exposure or centring chosen will be correct. This is of particular importance before such a procedure as aortography, for when an entire series of films has to be repeated, the increase in irradiation of the patient is considerable. But there are hardly any examinations where the quality of the final result is not improved by the use of the preliminary film.

In all, it will be seen that the radiologist must consider a very wide variety of measures if he is to maintain dosage to his patients at the lowest possible level.

APPENDIX

1. Polyfoam mattresses, Polyslip mattress covers and other items of foam plastic: Leslie's Ltd., (Polyfoam Division), Green Pond Road, Walthamstow, London, E.17.
2. Clysodrast: Lewis's Laboratories, Lavender Walk, Leeds 9. Veripaque (dihydroxyphenylisatin): Bayer Products Co., Surbiton, Surrey.
3. Bisacodyl: (Dulcolax), 5 mg. tablets are given to the patient in envelopes on which is printed:—

'The Tablets'

Take two for the $\frac{\text{night}}{\text{two nights}}$ preceding the X-ray examination, that is:—

2 tablets on....................at....................

2 tablets on...................at....................

On the day before your examination avoid eating green vegetables, bread, potatoes and fruit.

References

SCOTT-HARDEN, W.G., PHILP, L. D. and MOULE, B. (1967) *Br. J. Radiol.*, **40**, 15
WILD, A. A. and EVANS, J. (1968) *Br. med. J.*, **2**, 607

CHAPTER 2

SEDATION FOR RADIOLOGICAL PROCEDURES
(*Stanley Feldman, F.F.A.R.C.S.*)

WHEN preparing a patient for a radiological investigation it is necessary to consider whether sedation will be required and what form it should take.

There can be no simple formula to provide suitable sedation for all patients on every occasion. Each patient must be considered as an individual and his needs assessed from observation of his personality and his response to preliminary experiences in the X-ray department. The drug or a combination of drugs to suit both the patient and the particular radiological procedure to be performed should then be selected. When it is felt that suitable sedation cannot be achieved by safe dosages of the drugs available the advisability of performing the investigation under general anaesthesia must be considered.

As mentioned in Chapter 1, it is important to explain the procedure to the patient. Most patients assume all radiological examinations to be simple, quick and painless and may interpret the first uncomfortable sensation as evidence that something is wrong either with the investigation, or with themselves. Previous experience will help patients to relax and co-operate.

A good night's sleep is desirable before any major investigation. Patients who have slept badly are often tense and hypersensitive. Although anxiety is the usual cause of the insomnia there is little doubt that the sleeplessness itself produces further anxiety. In a survey of patients undergoing some form of manipulation or operative procedure it was found that nearly one-half of the patients failed to sleep adequately the previous night in spite of the routine administration of hypnotics. An increase in the dose of hypnotic or a change of drug will often ensure a good night's sleep for the patient.

Drugs

The conventional premedicant drugs that have been used in the preparation of these patients include

1. Opiates and synthetic analgesics.
2. Hypnotics.
3. Phenothiazine derivatives.
4. Parasympatholytic agents.

In addition (5) Tranquillising drugs and (6) 'Ataractic' agents can be used to obtain a well sedated, co-operative patient for some procedures.

The ideal premedication would produce a patient who was fully co-operative yet easily distracted and free from anxiety. He would be analgesic and insensitive to discomfort yet in full control of all his reflex activity. He would be relaxed and tranquil, yet free from such unpleasant sensations as dizziness on movement and he would be able to be raised from the lying to the standing position without producing hypotension. Unfortunately, it is not always possible to achieve this perfect state in patients by the use of drugs. Drugs which produce relaxation and distraction in a patient tend to lessen the patient's co-operation. Analgesics usually depress the protective reflexes and drugs which are useful in producing a feeling of relaxation and tranquillity often make the patient very sensitive to postural hypotension.

The greater the dose of drug the more frequently is the desired effect produced, but unfortunately a higher incidence of severe unpleasant side effects invariably occurs. Frequently the best that can be safely achieved by the use of sedative drugs is a compromise which merely alleviates the more unpleasant features of the examination.

Opiates and synthetic analgesics.—If the proposed procedure is likely to produce physical pain an analgesic should be administered as a premedication. Analgesics of all types are more effective in preventing pain than relieving established pain and should therefore be given some time before the proposed procedure.

All the opiates and the synthetic analgesics produce a depression of the respiratory minute volume and diminished sensitivity of the respiratory centre. They tend to lower the blood pressure and cause a depression of protective reflexes. In most patients these effects are of little importance, but in the feeble, the elderly, the hypovolaemic or the severely hypertensive patient they may be of sufficient extent to jeopardise the patient's life. Opiates and the synthetic analgesics should be given with extreme caution to patients with impaired pulmonary function from whatever cause and to patients with raised intracranial pressure since the resultant depression of respiration may prove fatal. Advanced liver disease will increase the sensitivity to opiates and the duration of action of these drugs will be prolonged.

For premedication, the opiates, such as papaveretum (Omnopon) and heroin, have the considerable advantage over the synthetic analgesics (e.g. pethidine) that they produce euphoria. Pethidine occasionally produces dysphoric symptoms such as dizziness, disorientation, light-headedness and nausea. It is the euphoria rather than the analgesia produced by the opiates that makes them the best agents available for preoperative sedation. 20 mg. of papaveretum (in a 70 Kg. adult) will usually produce a relaxed carefree individual who will sleep if undisturbed, but who can also co-operate sufficiently for the performance of most procedures. A higher dose will produce a greater incidence of well sedated patients but it will also produce a higher frequency of dangerous and unpleasant effects. It is better to supplement the opiate with another drug if a greater effect is required, or if the patient is especially anxious or the procedure particularly uncomfortable. A small dose of a phenothiazine often successfully augments the hypnotic action of opiates. The effect of this combination is usually very good; however the result may be difficult to predict.

Infants and children tolerate opiates well, but they are very susceptible to respiratory depression. Providing there is no pre-existing respiratory difficulty there is no reason why opiates should not be given to children in a similar dose/body-weight relationship as adults.

Hypnotics.—The hypnotics, especially the barbiturates, have proved disappointing agents for preoperative sedation in adults. After treatment with large doses of barbiturates patients behave as if recovering from a general anaesthetic, they are often confused, disorientated and unco-operative. Smaller doses of barbiturates have relatively little effect although it has been claimed by Beecher that 90 mg. of pentobarbitone produces analgesia equivalent to 10 mg. of morphia.

Infants and young children present special problems. They are easily frightened by the unfamiliar surroundings of the X-ray department and re-assuring words will have little effect on a child under the age of 18 months. These patients are best sedated by barbiturates and, ideally, they should be very sleepy or asleep, unless their active co-operation is essential. The difficulty lies in judging the time of the premedication accurately, especially if it is administered by mouth. It is best to give the barbiturate whilst the patient is in a quiet room near the X-ray department; the investigation is started as soon as he has fallen asleep. If the investigation is to last more than one hour the premedication invariably wears off too early and the radiologist is left trying to comfort a disorientated, frightened child. Under these circumstances either an alternative premedication (i.e. promethazine and pethidine) or a general anaesthetic is to be preferred.

The Phenothiazine drugs.—These drugs are unsatisfactory for premedication when used alone. They are extremely useful in potentiating certain actions of the opiates, the synthetic analgesics and the hypnotics, especially as they are relatively non-toxic and have a very low incidence of unpleasant or dangerous side effects.

Promethazine (Phenergan) has well marked hypnotic properties as well as being a potent antihistaminic. It is very useful in supplementing the hypnotic effects of the opiates, synthetic analgesics and the barbiturates and is claimed to have useful anti-emetic properties. Side effects are infrequent although when high doses are used somnolence for several hours may be produced. There is some evidence that like other phenothiazine drugs, promethazine actually lessens the effectiveness of analgesics in relieving pain. The usual dose of promethazine for a 70 Kg. adult is 50 mg.; it is most effective if administered with either pethidine or papaveretum. Promethazine is a useful hypnotic for infants and children in a dose of ½ to 1 mg./lb. body weight.

Trimeprazine (Vallergan).—This agent is recommended as an oral sedative for children. It can be used alone to produce sleepiness but is more effective if given with pethidine or papaveretum to increase somnolence. 0.5–1 mg./lb. body weight produces sleepiness in about 30 minutes. Like promethazine this drug is relatively free from side effects and the sleepiness it produces is associated with less disorientation than that produced by barbiturates.

Chlorpromazine (Largactil).—Chlorpromazine has been widely used to supplement the analgesic effects of opiates in the past. Although still used for its tranquillising and anti-emetic properties it has the disadvantage of frequently causing tachycardia, hypotension and pallor in the patient. Chlorpromazine is especially likely to produce a fall in blood pressure if administered together with a potent analgesic. The hypotensive effect is exacerbated by movement, particularly by suddenly assuming the upright posture. It should always be used with caution or avoided altogether if movement of the patient will be necessary during the radiological procedure. The usual dose for a 70 Kg. adult is 25 mg.

Perphenazine (Fentazin).—This is a very potent anti-emetic agent; it also possesses considerable tranquillising properties. 5 mg. of perphenazine will potentiate the mental relaxation produced by the opiates and synthetic analgesics. It is useful by itself for lessening anxiety in agitated patients.

Promazine (Sparine) is another useful drug which may be used, like perphenazine, either alone (50 mg. for an adult) or to potentiate the psychic sedation produced by other agents. Promazine, like chlorpromazine, may cause postural hypotension.

Parasympatholytic agents.—*Atropine and Hyoscine.*—One of these agents should be used whenever the proposed radiological manoeuvre involves the possibility of evoking a reflex vagotonic state, such as manipulation of the carotid artery for arteriography, or thoracic aortography. These procedures may produce alarming bradycardia unless atropine or hyoscine is given to the patient. Moreover, without these agents there is always a slight increase in the risk of cardiac arrest.

Hyoscine 0·4 mg. is also useful for procedures that may have to be repeated frequently as it has a central sedative action and produces a considerable amnesia for all but the most unpleasant experiences. Both atropine and hyoscine cause an unpleasantly dry throat in most patients which persists for up to 6 hours.

Tranquillising agents.—Diazepam (Valium) is the most versatile of these agents. 5–10 mg. given orally 1 to 2 hours before the procedure renders the patient more susceptible to the euphoria produced by opiate premedication, especially in the agitated patient. It can be given as an intramuscular injection to supplement the premedication, although it may cause some pain at the site of injection. Its greatest use is by the intravenous route. Given 5 minutes before the investigation 5–10 mg. intravenously will render the patient relaxed and sleepy but amenable to instruction. It does not have any analgesic properties, short of a sleep dose, and *must therefore be supplemented* if the procedure is likely to be painful.

Ataractic drugs.—Droperidol (Droleptan). This agent produces a state of chemical hypnosis. In a dose of 5 mg. i.-v. or 5 to 10 mg. i.–m. it produces a detached state of mind in the patient. Some patients find this loss of volition disturbing and it is better used in combination with a potent analgesic, such as phenoperidine (Operidine) 0·5 to 1·0 mg. or pethidine 50 to 100 mg. The combination of these drugs produces a virtually analgesic patient who is co-operative and in some instances it induces a virtual trance-like state in the patient. Phenoperidine, like pethidine, may produce respiratory depression and should be used with caution in elderly patients and in patients with respiratory disease. Heart disease is not a contra-indication to the use of these drugs.

Some suggested premedications

Children

1. For short procedures in which sleep is desired:—
   Rectal thiopentone 1 G. per 50 lb. body weight administered 1 hour before the procedure.
2. When sleepiness is desirable but co-operation may be required, or for longer procedures:—
   Pethidine 0·7 mg. per lb. body weight with promethazine 0·5 mg. per lb. body weight given intramuscularly 45 minutes before the procedure.

Adults

In adults papaveretum forms the best basis for premedication; a good sedative for an average 70 Kg. adult is:—
   Papaveretum 20 mg. and hyoscine 0·4 mg. given i-m. 45 minutes before the examination.

1. If procedure is prolonged but not painful and sleepiness is desired
   (a) Add 10 mg.–20 mg. diazepam 1 hour preoperatively in addition to normal premedication, or
   (b) give 10 mg. diazepam i.v. just before procedure, or
   (c) give a suitable phenothiazine 30 mins. preoperatively—perphenazine 5 mg. or promethazine 50 mg.
2. If procedure is painful
   give 5 mg. droperidol with 0·5 mg. phenoperidine intravenously 5–10 mins. before procedure.
3. If patient to be tilted frequently.
   reduce premedication dose of papaveretum to 10 mg. and supplement with perphenazine 5 mg. or diazepam 10 mg.

Should further sedation be needed during the procedure the dose of papaveretum may be repeated two hours after the time when it was first given (either intramuscularly or intravenously). The total dose of promethazine should be less than 1 mg./Kg. or some patients suffer from jerky, twitching movements. The maximum dose of diazepam should be 15 mg. if prolonged somnolence is to be avoided.

CHAPTER 3

EMERGENCIES IN THE X-RAY DEPARTMENT
(*Prepared with the assistance of Stanley Feldman, F.F.A.R.C.S.*)

EMERGENCIES are rare in X-ray departments but preparedness is vital so that when they do occur they may be treated effectively. Everyone who works in X-ray rooms should be familiar with simple techniques of resuscitation and should know what may be expected of them in helping to treat a collapsed patient. A warning system is needed so that help can be summoned quickly to any room where it is required. It is also best to have an 'emergency trolley' carrying all the drugs and equipment which may be needed (see Appendix). The trolley should be readily available, being kept in a set place known to members of the department. Failing this optimal state of readiness there is a certain basic minimum requirement of drugs and equipment also set out in the Appendix.

Reactions to Contrast Media[1]

The nature of reactions to contrast media is not fully understood although some are clearly allergic and the major reactions are anaphylactoid in type, the effects resembling those of histamine release (Mann, 1961). Most reactions are minor, such as arm pain, metallic taste in the mouth, giddiness, a feeling of warmth or tingling, coughing, nausea and vomiting. Such reactions usually pass off quickly and reassurance is all that is necessary. Headache is sometimes attributable to contrast media but is more often due to fluid deprivation.

Minor allergic reactions such as sneezing, rhinorrhoea, lacrimation, pruritus or urticarial rashes are much less common but are also relatively benign and will respond to antihistamines if necessary. It is the severe reaction which causes real danger and for which swift treatment is so important. Most such reactions occur within five minutes of injection and the great majority within thirty minutes, so that a doctor should be at hand for this period whenever an injection of contrast medium has been given.

Although it has been stated (Mann, 1961) that a second injection of contrast medium in angiocardiography involves a considerable increase in

---

[1] 1. This section, of necessity, draws largely from the published work mentioned in the references. Particular indebtedness is felt to the work of Pendergrass and his co-workers (1958) and Ansell (1966, 1968) who have studied this subject so carefully.

2. Not all reactions after injections are due to the contrast medium. Unrelated disease within the patient may simulate a severe reaction, e.g. pulmonary embolus (Finby *et al.*, 1958) or internal haemorrhage (Counts *et al.*, 1957). Minor reactions may be due to pyrogens and other contaminants, especially when syringes or needles are not properly washed or sterilised. Simple vaso-vagal fainting may occur from fear of the needle or as a reaction to the puncture itself. Very rarely a hypertensive attack due to a phaeochromocytoma may occur, resembling a reaction to contrast. This is further discussed later in this chapter.

the risk of producing a reaction, there is little to suggest that this is so in other types of investigation nor does vomiting after contrast usually recur if a second injection is given.

Prophylaxis.—All test procedures are unreliable and it is a mistake to depend upon them. The most that can be recommended for regular use is the initial injection of 1 ml. of contrast followed by a pause of sixty seconds before injecting the full amount. There is, incidentally, no evidence that the rate of injection has any effect on the frequency of major reactions. Our own experience has been that this also applies to minor reactions when injecting for intravenous urography and we make such injections rapidly. However, this point is more debatable and there is some variation with different media, e.g. Biligrafin should be injected over at least three minutes and when given as a drip-infusion causes fewer side-effects.

Probably the most useful prophylactic measure is to enquire for a history of allergy or previous reaction to contrast, although even this will only give warning of a proportion of reactions. Questioning should be thorough —not merely 'do you suffer from hay-fever, nettlerash or asthma', but 'are you allergic to anything; have you ever reacted to any medicines, tablets or injections'. It is important to establish a routine whereby a given member of Staff, whether radiographer or nurse, asks these questions as soon as possible after the patient arrives.

In cases where there is a history of simple allergy, mild asthma or slight allergic reaction to previous injection of contrast it is advisable if only for medico-legal reasons to give an antihistamine. Chlorpheniramine (Piriton) is quick-acting, effective and not too hypnotic. 10 mg. may be given intramuscularly thirty minutes before or intravenously ten minutes before the contrast is injected. Out-patients should be warned that they may become a little sleepy since this may affect their travelling home. Promethazine (Phenergan) is a more powerful antihistamine but has a stronger hypnotic effect, and is less suitable for out-patients. 25–50 mg. are given intravenously or intramuscularly.

When there is a history of severe allergy, asthma or previous reaction to a contrast medium, the following measures should be taken[1]:—

1. The need for the examination is critically reviewed and, if necessary, discussed with the clinicians concerned. It may be possible to obtain the desired information in some other way, e.g. by isotope renography.
2. If it is decided to proceed, an antihistamine[2] is given and intravenous steroids are given at least thirty minutes before the contrast.
3. An intravenous catheter (e.g. Plextrocan) is employed for injection and is left in the vein for 30 minutes. The radiologist should be on hand during this period and the resuscitation trolley must be available.

[1] In severely sensitive patients it is probably not safe to assume that a retrograde pyelogram can be carried out instead of a urogram. Tests using radio-iodine tagged contrast media have shown quite considerable uptake into the circulation from the renal pelvis.
[2] Most antihistamines cause drowsiness and the patient must be given time to recover from this before being allowed home unescorted, particularly before driving a car.

4. When the nature of the contrast medium to which a reaction occurred previously is known, a different type of medium is administered.

Premonitory Symptoms.—Although nausea and vomiting are common as a minor reaction they occasionally herald a severe disturbance and the radiologist is wise to assure himself that these complaints are subsiding before leaving the patient. *Restlessness* is another symptom which is usually unimportant but may be part of a reaction, and the patient with restlessness should be watched until it improves. Unusual pains and other bizarre sensations have also been reported during the early stages of severe reactions.

The Types of Severe Reaction.—These may be classified for therapeutic convenience as major allergic reactions, respiratory difficulty and circulatory collapse. In fact, the underlying mechanism, i.e. release of histamine and kindred substances is similar in all these cases and the different 'types of reaction' result from individual responses by different patients. For this reason the symptomatology is not always clear cut. However, this classification draws attention to those effects which endanger life and must therefore be treated.

A. *Major allergic reactions.*—These are angioneurotic oedema, giant urticaria and laryngeal oedema. The latter may occur as part of an attack of angioneurotic oedema, or as an isolated manifestation; it is recognised by stridor, difficulty with inspiration and inspiratory recession of the tissues of the neck and chest.

B. *Respiratory difficulty.*—This is only occasionally due to laryngeal oedema; much more commonly it is due to bronchospasm or to pulmonary oedema. Bronchospasm usually occurs in a known asthmatic, the patient showing typical wheezing, with difficulty in expiration; less often it occurs in chronic bronchitics. Pulmonary oedema has occurred relatively commonly among reported cases of severe reaction and is to be feared as a dangerous type of reaction which in some cases is accompanied by a state of circulatory collapse. The patient rapidly becomes dyspnoeic with cyanosis and frothy sputum.

C. *Circulatory collapse.*—This presents with restlessness, confusion, pallor, sweating and air-hunger. The blood pressure falls and the patient may lose consciousness. Cyanosis develops, pupils dilate and convulsions due to anoxaemia sometimes occur.

Treatment of the Severe Reactions

A. General measures—

1. If untoward symptoms occur before the injection of contrast ends, discontinue injection but *leave the needle in the vein*. Should it appear that a severe reaction may be developing, Betnesol 4–8 mg. is made ready and can then be given if necessary.
2. If the reaction begins after removal of the needle a rubber tourniquet may be applied to the arm to engorge the veins.

3. The signal calling for assistance is given.
4. A finger is kept on the pulse for as much of the time as possible.

As these major reactions are all allergic or anaphylactoid the two drugs which are of most value are adrenaline and cortico-steroids. The main theoretical drawback of steroids is that even hydrocortisone or predniso-lone given intravenously are thought to take some time, i.e. ten to twenty minutes, before exerting their full effect. Notwithstanding this, steroids have proved effective in the therapy of such cases (Besterman *et al.*, 1956; Wright, 1959) and should always be given. Betnesol has the advantage that no mixing is needed. The protective effect may wear off quite rapidly and repeated injection may be required.

B. Specific measures—

Major allergy.—

1. Give Betnesol 4–8 mg. intravenously.
2. Give also adrenaline 1:1,000 0·5 ml. subcutaneously; repeat after 10 and 20 minutes if no improvement has occurred.
3. Give promethazine (Phenergan) 50 mg. intravenously if it has not already been given.

In most cases these measures will control a major allergic reaction—continued dosage of antihistamine and steroids may be needed for up to 24 hours or more, particularly if renal failure causes delayed excretion[1].

4. Oxygen may be given if breathing becomes difficult. Should laryngeal oedema worsen in spite of this treatment, measures to establish an airway will be needed. A gradual worsening will allow time for a surgical colleague to be called to carry out tracheostomy. With rapidly developing laryngeal obstruction the radiologist may have to relieve it himself. For this purpose—remembering that the obstruc-tion is likely to be of short duration—laryngostomy is recom-mended as simpler and quicker for the unskilled. For details see below (p. 22).

Respiratory difficulty.—Although treatment is somewhat similar for the two causative conditions one should try to distinguish between broncho-spasm and pulmonary oedema, as there are some differences in their therapy.

*Bronchospasm.—*The asthmatic patient may start with his own usual treatment, but if this fails:—

1. Give isopenaline 0·5–1·0 ml. 1:400 solution intravenously; repeat after 10 and 20 minutes if necessary; then, if the attack persists (or sooner if it worsens)—
2. Give Betnesol 4–8 mg. intravenously.
3. Give aminophylline 0·5 G. intravenously slowly.
4. Give oxygen as needed.

[1] In such cases dialysis is advisable to elminate the contrast as rapidly as possible.

Pulmonary oedema

1. Give oxygen, preferably under positive pressure (p. 22) immediately. With sufficient pressure it may be possible to reverse the pulmonary oedema. Should the accompanying drug therapy fail, tracheal intubation will make positive pressure oxygen therapy much more effective. The patient is propped up unless he is in circulatory collapse.
2. (*a*) Give adrenaline 1:1,000 0·5 ml. subcutaneously. This drug may seem surprising treatment for pulmonary oedema but it is specific therapy in anaphylaxis (Pendergrass *et al.*, 1958). Should there be any suspicion that the pulmonary oedema is due to heart disease adrenaline is contra-indicated.
   (*b*) Give Betnesol 4–8 mg. intravenously.
   (*c*) Give aminophylline 0·5 G. intravenously.
   (*d*) Give frusemide (Lasix) 20 mg. intravenously.

Circulatory collapse.—The aim of treatment is to *maintain an adequate circulation of oxygenated blood.*

1. Lower the head of the couch and give the signal for assistance.
2. Make sure the airway is adequate and give oxygen. An assistant should take this over and should monitor the carotid pulse.
3. Give adrenaline 1:1,000 0·5 ml. intramuscularly and massage the site. If there is severe shock give half the dose intravenously.
4. Follow this immediately by Betnesol 8 mg. intravenously using the jugular or femoral vein if necessary.
5. Measure the blood pressure. A blood pressure of more than 60 mm. Hg. systolic is adequate in the head-down position provided the patient is fully oxygenated.
6. With a blood pressure below 60 mm. Hg. systolic give mephentermine 5 mg. intravenously every 5 minutes. Metaraminol (Aramine) can also be used; it is best given in consultation with an anaesthetist and preferably in an intravenous saline drip (50–100 mg. to 500 ml. of normal saline or 5 % dextrose).
7. Convulsions are a sign of inadequate oxygen supply to the brain and call for more effective administration of oxygen with maintenance of an adequate blood pressure. Sedatives should *not* be given until these requirements have been fully met; a short-acting barbiturate such as thiopentone may then be given with caution.
8. Cardiac arrest may occur. Its treatment is described below.

Phaeochromocytoma.—The effects of a phaeochromocytoma are due to release of noradrenaline and adrenaline, intermittently or continuously. Noradrenaline results in hypertension while adrenaline gives rise to tachycardia, sweating, palpitations, peripheral vasoconstriction and sometimes a sense of anxiety or fear. In most cases symptoms due to both noradrenaline and adrenaline are found; occasionally one or the other predominates. The effects of noradrenaline are counteracted by so-called α-blockers. Of these the most rapidly acting is phentolamine (Rogitine) which has an immediate effect when given intravenously, while the longest acting is phenoxybenzamine which is used for the pre-operative control of symptoms when release of noradrenaline is continuous. At the present the most effective β-blockers are propanolol (Inderal) and practolol. To control symptoms thought to be mainly due to noradrenaline, phentolamine is given intravenously, 3 mg. initially followed by additional amounts of 1 mg. every 2 minutes up to 10 mg. or until the blood pressure is controlled. Occasionally the resulting drop in blood pressure is catastrophic and administration of noradrenaline, 0·5–0·75 ml. of 1:10,000 solution, is needed. It is advisable to give 0·25 ml. and then further increments of 0·1 ml. every minute until the B.P. is normal. Lowering the head of the table and administration of oxygen may be advisable as well. (If noradrenaline is not available massage of the abdomen may cause release of endogenous noradrenaline). A continued fall in blood pressure may make it advisable to start a saline drip with added noradrenaline.

The symptoms of adrenaline release, notably cardiac irregularities and tachycardia, are controlled by slow injection of propanolol 1 mg. a minute up to 5 mg. Even in cases with predominant 'adrenaline' symptoms it is advisable to give intravenous phentolamine 2–5 mg. as well, with careful watching of the blood pressure.

The outlook in phaeochromocytoma is considerably improved and the need for emergency measures as described above are diminished if the effects of catecholamine release are controlled *before investigation* and operation. The following routine is therefore suggested in suspected cases:

1. Whenever possible the diagnosis should be established by estimation of urinary catecholamines before radiological investigation.
2. The suprarenal areas should be examined with a high dose of contrast (e.g. 100 ml. Conray 420) and immediate tomography. This is sometimes sufficient to show a suprarenal tumour, making further investigation unnecessary.
3. If presacral pneumography or suprarenal venography or arteriography are still needed the patient's symptoms should be controlled by α-blockers, e.g. phenoxybenzamine and, where appropriate, β-blockade (propanolol) for several days beforehand. Before the investigation an intravenous cannula is inserted and phentolamine, propanolol and noradrenaline are available. The blood pressure and heart rate are regularly measured and the patient's general condition is carefully watched. If a reaction does occur it should be managed as outlined above; the patient is transferred to the care of a physician as soon as possible.

Occasionally symptoms due to an unsuspected phaeochromocytoma may occur during the course of a radiological investigation. This may give rise to considerable difficulty in diagnosis particularly when contrast has been injected since some of the symptoms and signs—dyspnoea, palpitations, tightness in the chest, headache, 'goose-flesh' and sweating—are equally attributable to the contrast. Measurement of the blood pressure may make a distinction possible since it may be elevated in phaeochromocytoma and for this reason the radiologist should try to bear the possibility in mind when confronted by puzzling reactions.

Reactions to local anaesthetics.—Administration of any local anaesthetic carries a risk of producing toxic effects and it is advisable to make a routine of starving the patient for 4–6 hours before the use of a local anaesthetic. Loss of consciousness is then less likely to be accompanied by aspiration of vomit. In addition it is most important to have oxygen available for administration and a local anaesthetic should *never* be given without oxygen being ready.

The toxic dose of local anaesthetic varies inversely as the square of the concentration, i.e. the higher concentrations need a smaller amount to produce a toxic effect. Lignocaine is a safe local anaesthetic which has been extensively used and is normally the first choice. Prilocaine (Citanest) is a newer local anaesthetic claimed to have a lower toxicity and is therefore useful for topical anaesthesia in the bronchi. It may cause apparent cyanosis due to methaemoglobinaemia. Table X in the General Appendix gives the manufacturer's recommendations on maximum dosage. Excessive use as a topical anaesthetic or accidental injection into a vessel may give rise to symptoms:—

1. *Sedative effects.*—These are the most common sequelae of lignocaine overdosage, the patient becoming increasingly drowsy and possibly complaining of dizziness. He should be allowed to sleep off the overdose under supervision. The most satisfactory arrangement is that he should be given routine post-anaesthetic observation in a ward—i.e. quarter-hourly pulse and half-hourly blood pressure—until he is awake. He should at no stage be given fluids, such as tea, in an attempt to wake him, for if his level of consciousness later falls there is an increased risk of aspiration of vomit.

2. *Convulsive effects.*—These are much less common with lignocaine than with other local anaesthetics. They vary from twitchings to full convulsions, and result from a sedative action which selectively depresses parts of the cortex, releasing other parts; it is therefore a mistake to give large amounts of sedative in treatment. Persistent twitchings can be controlled by giving thiopentone intravenously very slowly until the patient falls asleep. When convulsions are frequent, however, relaxant drugs should also be given. Scoline 100 mg. intravenously (or intramuscularly if the convulsions make intravenous injection impossible) will paralyse the patient for a short time and make possible insertion of an airway and insufflation with oxygen. An anaesthetist may then be called to take over care of the patient.

3. *Hypotension.*—This is treated with pressor drugs and oxygen as described above under 'Circulatory Collapse'.

4. *Cardiac arrest.*—The treatment is described later in this chapter.

Amethocaine is less frequently the cause of sleepiness and is more likely to cause nausea, hypotension, convulsions and cardiac arrest. Thiopentone is the main drug used in treatment.

Treatment of gas embolism.—Gas embolism may be venous—that is on the right side of the heart—or arterial. The latter is uncommon after radiological procedures but may result from:—

1. Direct injection via an arterial catheter or needle.
2. Passage across an intracardiac shunt following intravenous injection.

No useful treatment is available for this type of gas embolism.

Venous embolisation with gas is also an unusual complication. It most often occurs during procedures in which large amounts of air or oxygen are injected into the peritoneal cavity or retroperitoneal tissues (i.e. pneumoperitoneum or presacral pneumography). Quite large amounts of gas may be injected intravenously with the patient lying on the left side and the gas can collect in the right atrium without necessarily obstructing the circulation or causing symptoms (Stauffer *et al.*, 1956). When the patient is put on his back or right side the gas immediately collects in and obstructs the outflow tract of the right ventricle. The patient then collapses with dyspnoea, cyanosis, rapid fall in blood pressure, loss of consciousness and a millwheel murmur over the heart, sometimes audible without a stethoscope.

Precautions to avoid air-embolism are described in the relevant sections. The routine for treatment is:—

1. Stop injection of the gas.
2. Turn the patient on his left side and lower the head—if the table does not tilt put a pillow or other support under the buttocks.
3. Give oxygen by face mask under positive pressure.
4. Feel the carotid pulse. If the heart stops open cardiac massage is performed. Before massaging the exposed heart gas should be aspirated from the right atrium with the largest available needle and syringe. Should there be no large syringe available, a large needle is inserted and air squeezed gently out through it.

Electric shock.—The measures to be taken are:—

1. Turn off the electricity at the mains. The main switch in every X-ray room should be painted a bright colour with a similarly coloured arrow on the wall above it.
2. If it is not possible to turn off the electricity the subject is pushed or pulled away from the live terminal using an insulating material. It may, of course, be easier to push the terminal from the patient. A wooden broom is a suitable instrument for pushing or pulling. A blanket can be passed round a patient or the limb in contact to pull it away from the terminal. A tightly folded blanket or rolled-up pillow are suitable for pushing.
3. Mouth-to-mouth respiration or oxygen administration must be

B

started if breathing has stopped (p. 23). Cardian massage and defibrillation may also be needed (pp. 24 and 26).

Administration of oxygen.—The importance of oxygen in such conditions as those we are discussing cannot be over-emphasised. Without adequate oxygenation other measures are valueless. Moreover it is not always sufficient to give oxygen at rates which correspond with normal breathing; when a patient suffers from circulatory collapse or cardiac arrest he develops a considerable acidosis and needs to be hyperventilated with oxygen. This also applies during recovery.

Conscious patients can be given oxygen by any of the normal methods. So may unconscious patients who are breathing spontaneously but such patients should be kept head-down or on one side because of the risk of aspiration of vomit. Those unconscious patients with respiratory depression or apnoea need some form of positive pressure administration. This is most simply provided by an anaesthetic mask and bag which should be the standard fitting on all oxygen cylinders. (To provide positive pressure administration for patients who are breathing spontaneously the oxygen is given at a rate which will keep the reservoir bag distended.) A box or bag attached to the cylinder stand should contain a child-size face mask and Guedel airways sizes 1, 2 and 3. Use of the face mask and airway should be discussed with an anaesthetist colleague and demonstrations given

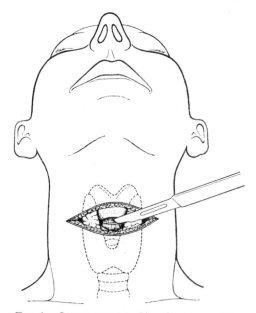

FIG. 1.—Laryngostomy. Showing the position of the head and the site of incision.

to members of staff who may be called upon to assist. In hospitals where help from an anaesthetist would not be quickly available the radiologist may wish to learn to pass an endotracheal tube and this too may be discussed with an anaesthetist.

Laryngostomy.—This has been recommended (Abbey, 1960) for the relief of acute laryngeal obstruction in adults. It is simpler and quicker than tracheostomy, the scar is smaller and the area used less vascular. It is not suitable as a long-term measure but in the relief of an emergency, such as glottal oedema, which is unlikely to last for long, it has much to recommend it.

Method (Fig. 1).—The head is fully extended, the thumb and middle finger of the left hand grasp the thyroid and cricoid cartileges to steady them. The index finger locates the crico-thyroid 'gap' and a scalpel or knife in the right hand is used to make a 2 cm. transverse cut over the crico-thyroid membrane, exposing it. The membrane is divided with the next cut. If a laryngostomy tube is available it is now inserted or a small endo-tracheal tube with a safety pin through it may be used. Failing this the handle of the scalpel is put in—directing it downward—and twisted so as to open the space enough to create an airway.

Mouth-to-mouth artificial respiration.—As well as understanding how to give oxygen, members of a department should be trained in the technique of mouth-to-mouth artificial respiration in case there is any delay in obtaining oxygen. The technique is best learned by watching demonstrations or films; the details are as follows:—

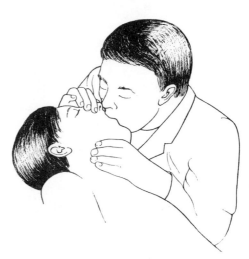

1. If there is any reason to suspect the presence of foreign matter such as vomit in the throat a finger is swept round the back of the throat to clear it. It may be necessary to turn the patient prone and squeeze his chest vigorously if the larynx is obstructed. False teeth are *not* removed, but should be properly in place.

   Fig. 2.—Mouth-to-mouth respiration. Position of the patient and operator during inflation of the lungs. The operator's hand pinches the nose.

2. The head is extended and the jaw pulled well forward into a position of prog-nathism. The jaw may either be lifted from behind the angle of the mandible or a (padded) thumb put inside the mouth to grasp it. If an airway is available it is inserted—either a conventional airway or the double airway designed for mouth-to-mouth respiration.

3. A handkerchief is laid over the mouth or nose, whichever is to be used, and the operator takes a deep breath and lays his mouth against it. (In children both the mouth and nose may be covered; in adults the nose is satisfactory when it is not blocked.) The position of the fingers to seal off the mouth or the nose and to keep the jaw pulled forward is shown in Fig. 2. (In children one hand presses on the abdomen to prevent gastric distension.) The use of a handkerchief is not a necessary part of the procedure but helps to make it less repugnant to some.

4. The operator blows—gently for children, forcefully for adults—and watches the chest rise. He then removes his mouth and allows expiration to occur (Fig. 3). It is important to watch the chest and make

sure it moves satisfactorily. If the airway is not adequate it must be cleared as soon as possible—see para. 1.

5. The cycle is repeated twenty times a minute, slowing the rate if one becomes dizzy or feels other symptoms of hyperventilation. In cases where no oxygen becomes available respiration is continued until the patient breathes adequately by himself—not merely in gasps but with normal breathing. Sometimes it may be necessary to start respiration again for a second or third period.

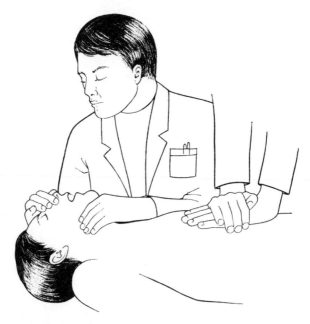

FIG. 3.—Mouth-to-mouth respiration and external cardiac massage. The operator raises his head to allow expiration to occur. The position of the hands for external cardiac massage is also shown.

Cardiac massage

Cardiac arrest is diagnosed by the disappearance of the carotid pulse. As soon as it is confirmed, treatment must start and assistance be called. Once the pulse is no longer palpable time must not be wasted in attempting to inject drugs of any kind—a useless procedure when there is no circulation. It must be emphasised that three to four minutes is the longest acceptable delay before instituting cardiac massage; after this brain damage is usually irreversible. *Cardiac massage by itself is valueless without adequate oxygenation* and, as in circulatory collapse, hyperventilation with oxygen is desirable to reverse the acidosis. (If oxygen is not available mouth-to-mouth respiration is administered.) The assistant who gives the oxygen should monitor the carotid pulse. The routine is as follows:—

1. Give oxygen—see p. 22.
2. Lower the head of the couch and raise the patient's legs vertically to increase the venous return, then bend them into the abdomen to compress the liver. This is sometimes enough to start the heart but is not continued unless an assistant becomes available to do it.

External cardiac massage.—The hard surface of an X-ray couch is a good resistance against which to press but a platform may be needed so that the operator can get above the patient. As external massage begins an assistant obtains a scalpel in case the chest has to be opened. The hands are placed one on top of the other and the heel of the underneath hand is pressed on to the lower sternum (Fig. 3). To compress the heart the operator pushes downward with a fairly strong jerk so as to depress the sternum about two inches. The pressure is rapidly released and the sternum moves up again. This cycle is repeated about sixty times a minute. On the occasions, unusual in hospital, when one operator has to perform external massage and mouth-to-mouth respiration one full insufflation with air is given after eight depressions of the sternum. An assistant feels the carotid region to see whether an effective pulse is produced; if not, internal massage will be needed. Should external cardiac compression fail to establish a palpable pulse within 3 minutes, then internal cardiac massage should be performed.

Internal cardiac massage.—The operator must not hesitate to open the chest because of lack of sterility; time spent in sterilising the skin and hands is so much time wasted. The incision is made below the left nipple and runs medially from the anterior axillary line to an inch from the outer border of the sternum (Fig. 4). (The site usually recommended is between the 4th and 5th ribs but, except in patients with pendulous breasts, under the nipple or under the breast will give the same exposure without losing time in counting ribs. Where the breasts are pendulous incision is made at this level above the breast.) At the medial end of the incision care is taken not to divide the internal mammary artery which lies just lateral to the sternum.

Any sign of bleeding is an indication that the heart has not stopped and other pulses as well as the carotid should be felt. When the chest is opened the lung may fall away and the pericardium may be visible. The ribs are wrenched apart manually, cracking the costal cartilages, until the heart is adequately exposed; if necessary the cartilages are cut. The exposed lung is now observed to make sure it moves well with oxygen insufflation, and the heart is inspected to see whether there is any pulsation.

At this stage it is usually more convenient to stand on the patient's left. If the heart shows no sign of movement it is intermittently compressed from behind against the sternum using the right hand; massage is attempted in this way for about fifteen seconds. Unless a good pulse can be produced, however, the pericardium must be opened.

If, on first inspection, the heart is seen to be fibrillating it is given a few sharp flicks; this may restore normal rhythm but otherwise massage is carried out.

When the pericardium has been opened the heart is squeezed and released rhythmically between the palms and palmar surfaces of the fingers; the right hand passes behind, the left in front; the *finger-tips must never be used.* In children and smaller patients massage may be one-handed, squeezing between the palmar surfaces of the fingers and the thenar eminence. This is not effective in larger patients. Regular compression is continued thus at about sixty beats a minute as long as it is necessary. It is tiring work and the hand positions will need to be altered and if possible an assistant take over for periods. As soon as possible an anaesthetist and a surgeon are called to take over the care of the patient.

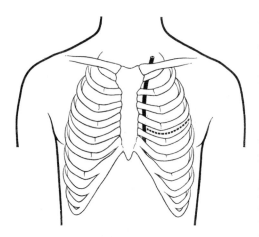

Fig. 4.—Internal cardiac massage. The site of incision. The line of the internal mammary artery is also shown.

Sodium bicarbonate

As already mentioned cardiac arrest produces a marked acidosis. This is only partly countered by adequate oxygenation and any case which has been in cardiac arrest for more than 30 seconds should be infused with 50–100 ml. of 8·4% sodium bicarbonate. Further infusion may be needed but should be controlled by estimations of the serum bicarbonate.

Defibrillation

In many cases of cardiac arrest there is ventricular fibrillation. If this persists after a period of effective cardiac massage defibrillation may be necessary. In some hospitals a trained team is available for cases of cardiac arrest and will undertake defibrillation. Radiologists who cannot look for such assistance should familiarise themselves with the use of the defibrillator and the electrocardiograph to control its application.

References

Reactions to contrast media

BESTERMANN, E. M. M., LEONARD, J. C., and WOOD, P. (1956). *Br. med. J.*, ii, 695.
COUNTS, R. W., MAGILL, G. B., and SHERMAN, R. S. (1957). *J. Am. med. Ass.*, **165**, 1134.
FINBY, H., EVANS, J. A., and STEINBERG, I. (1958). *Radiology*, **71**, 15.
MANN, M. R. (1961). *Proc. R. Soc. Med.*, **54**, 473.
PENDERGRASS, H. P., TONDREAU, R. L., PENDERGRASS, E. P., RITCHIE, D. J., HILDRETH, E. A., and ASKORITZ, S. I. (1958). *Radiology*, **71**, 1.

WEIGEN, J. F., and THOMAS, S. F. (1958). *Radiology*, **71**, 21.
WRIGHT, F. W. (1959). *Br. J. Radiol.*, **32**, 343.

Laryngostomy
ABBEY, P. (1960). *Lancet*, i, 573.

Gas embolism
STAUFFER, H. M., DURANT, T. M., and OPPENHEIMER, M. J. (1956). *Radiology*, **66**, 686.

Further reading
ANSELL, G. and ANSELL, A. (1964) *Br. J. Radiol.*, **37**, 881.
ANSELL, G. (1968). *Clin. Radiol.*, **19**, 175.
BARNHARD, H. J. and BARNHARD, F. M. (1968). *Radiology*, **91**, 74.

APPENDIX

Emergency equipment—basic minimum

1. *Oxygen cylinder* with 6–9 feet of tubing. Adult face mask and anaesthetic rebreathing bag. Attached to the cylinder stand a bag or box with:—
    Child size face-mask.
    Airways sizes 1, 2 and 3 (+ tube of lubricating jelly).
    Scalpel (for cardiac massage) in a sterile box.

2. *Syringes* (included on all trolleys or trays when contrast is to be injected):—

    $\left. \begin{array}{l} 2 \times 10 \text{ ml.} \\ 4 \times 2 \text{ ml.} \end{array} \right\}$ Disposable syringes.
    4 each disposable needles hypodermic size No. 1, 2 and 12.

3. *Drugs* (included on all trolleys or trays):—

Chlorpheniramine maleate (Piriton)	2×10 mg. ampoules.
Promethazine (Phenergan)	2×50 mg. ampoules.
Adrenaline 1 : 1000	2×1 ml. ampoules (keep away from light; change when pink).
Betnesol	2×4 mg. ampoules.
Isoprenaline 1 : 200 and 1 : 400 solution	2×100 mg. ampoules.
together with ampoule files.	2×1 ml, ampoules.

4. *Rubber tourniquet.*

Full emergency trolley

1. Oxygen cylinder, masks, etc., as above together with cuffed endotracheal tubes sizes 0–1–2–5–7–9, artery forceps, nylon syringe, Mackintosh laryngoscope.

2. Syringes, as above, together with:—

    2×20 ml. syringes
    $2 \times\ \ 5$ ml. syringes
    and a $2\frac{1}{2}$-inch, 17-gauge needle for drawing up.

3. Drugs, as above, together with:—

    Aminophylline 2×0.25 G. (in 10 ml.) ampoules.
    Mephentermine Sulphate (Mephine) 4×1 ml. ampoules (15 mg. each).
    Metaraminol bitartrate (Aramine) 1×10 ml. vial (contains 100 mg.).
    Thiopentone sodium 1×1 gm. ampoule (+ ampoule of water).
    Succinyl choline chloride (Scoline) 2×100 mg. ampoules.
    Phentolamine (Rogitine) 2×5 mg. ampoules.

Frusemide (Lasix) 2 × 20 mg. ampoules.
Noradrenaline (Levophed-Special) 1 : 10,000.

4. Other items:—

Intravenous giving set and 500 ml. bottle of normal saline. 150 ml. of 8·4% sodium bicarbonate.
Airways for mouth-to-mouth respiration (*P.P.*), sizes 1–2 and 2–3.
Tracheostomy pack.
Suction device (preferably from piped suction in all X-ray rooms) and catheters.
Sphygmomanometer and stethoscope.
Instruction sheet outlining the more important points about the treatment of reactions. An excellent example is the 'Notes on Radiological Emergencies', (1966) by G. Ansell, published by E. and S. Livingstone, Ltd., Edinburgh and London.

CHAPTER 4

ASEPSIS IN RADIOLOGICAL PROCEDURES
(prepared with the assistance of Professor B. W. Lacey)

RADIOLOGISTS tend to regard asepsis rather lightly when carrying out X-ray investigations. Normally their justification for this is twofold; first the difficulty of maintaining aseptic conditions in the average X-ray room, and second the alleged rarity of infection following the various types of procedure described in this book. Even when allowance is made for the indifference of one's memory for such events and for the lack of any critical follow-up, it is true that infection after any X-ray investigation is still surprisingly rare except after some examinations of the urinary tract. This, however, hardly excuses a careless routine for it is a tragedy if a patient is ever injured by a diagnostic procedure. A few notes are therefore included on 'sterility' in X-ray procedures. It may be remarked that there is much to be said when planning a new X-ray department for the inclusion of a 'procedures room' where a high level of asepsis may more easily be maintained during X-ray procedures. When an ordinary diagnostic room is used it is desirable to improve the level of asepsis as far as possible. The doors should have bolts to stop the unwanted entry of those not concerned in the investigation.

Scrubbing up.—In many centres the examinations described in this book are performed with the hands uncovered. As noted below, this is not the optimal approach but in any case the agents used for 'scrubbing up' should be both effective and lasting. For this reason there are especial advantages in the use of agents such as hexachlorophene or chlorhexidine (Hibitane) in the course of preparing the hands. Hexachlorophene is slow-acting and does not produce a rapid disinfection of the skin but its effect is prolonged and is probably enhanced by repeated use. The alternative methods recommended are:—

1. Wash for two minutes in running water using hexachlorophene soap. The nails (which should be kept short) and any areas of ingrained dirt are cleaned with a nail brush, but the hands and forearms generally are simply washed. After rinsing, the hands and forearms are dried on a sterile cloth or paper towel and then swabbed copiously with 0·5% chlorhexidine in 80% spirit.

2. Scrub-up using povidone-iodine (Betadine); the hands and forearms are wetted and the Betadine worked into a slight lather. The nails are cleaned with a scrubbing brush or orange stick, the hands and forearms rinsed and washing with Betadine is repeated continuing for a full two minutes. The hands are then rinsed and dried.

It is preferable to use nylon nail-brushes with plastic backs. Ideally these should be autoclaved and be available in a dispenser. If this is not possible they should be boiled once a day and be kept immersed in an efficient disinfectant, e.g. 1·5% Hycolin.

If, during the course of a procedure one accidentally touches an unsterile object it is normally sufficient to swab the hands thoroughly with Hibitane/spirit and then allow them to dry, but if there is reason to believe the object to be contaminated with infective material a fresh scrubbing is needed.

The use of gloves.—It is a common misapprehension that the use of gloves impairs the sensitivity of the fingers, making arterial puncture difficult. This is not so, and with a little experience there is no difficulty in arteriography when gloves are worn. Many other procedures, particularly those involving the urinary tract—micturating cystography, renal puncture or percutaneous nephrostomy—are made safer by wearing gloves and they are therefore included in the trolley settings. Another important reason for wearing gloves is to protect the wearer against infection from the patient. This aspect is most significant when carrying out procedures on patients from a dialysis unit because of the high risk of hepatitis. In such cases even simple venepuncture should be performed with gloves on and the gloves should be carefully removed and disposed of together with the syringes and needles which have been used.

Masks and gowns.—The wearing of masks is a widespread practice but it is doubtful whether it contributes directly to asepsis when performing punctures. Less common is the wearing of a sterile gown which we would recommend for any procedure where a lack of asepsis might have serious consequences. The main argument for doing this is that when wearing a protective apron it is very difficult to avoid brushing against and so contaminating sterile trolleys and sterile towels. Wearing a gown removes this source of potential contamination.

Handling needles.—The chief reason why sepsis is uncommon after most radiological procedures is that needles are used and the area of tissue exposed to possible infection is very small. It is, even so, a good practice always to hold the needle by the butt and to avoid touching the shaft—a practice which is, of course, standard in carrying out lumbar puncture.

Skin Disinfection.—As with preparation of the operator's hands, 'sterilisation' of the patient's skin can only be relative and means a reduction in the number of bacteria, particularly the vegetative forms. Before venepuncture, or puncture of very superficial arteries, swabbing with chlorhexidine/spirit and allowing to dry is adequate, provided the skin and underlying tissues are healthy. (Allowing the spirit to dry increases its effectiveness and prevents the stinging caused by puncture through 'wet' spirit.) When, however, a needle has to pass into much soft tissue, into a relatively avascular area, a joint space or the spinal theca, or through potentially contaminated skin—e.g. in the incontinent—or through devitalised tissues —e.g. in advanced diabetes—under all these circumstances skin disinfection should aim at killing spores as well as vegetative organisms. It is commonly argued that when puncturing vessels aseptic technique is of little importance because organisms introduced into the bloodstream are rapidly phagocytosed. This may be true but it must be remembered that the needle usually has to pass through some soft tissues and that arterial puncture almost always leads to haematoma formation and hence the provision of an excellent culture medium for some bacteria. For all but the simplest forms of puncture, therefore, the most reliable method of preparing the skin should be used. This is to paint the area of intended puncture with 1·5% iodine in 80% isopropyl alcohol, allow to dry, paint again and allow to dry. The iodine can then be washed off with spirit. Application of iodine should be careful and limited; if put on freely it may trickle off the area concerned and cause a 'burn', e.g. in the groin. By the time it has dried its disinfecting action is complete and washing off with spirit removes the risk of reaction in patients hypersensitive to iodine. The preparation with spirit may be extended over a much wider area than the region of the puncture.

Areas such as the vulva, urethral meatus, perineum and the mouths of sinuses are even more difficult to disinfect than normal skin. Disinfection is probably best achieved by washing thoroughly with domiphen-chlorhexidine solution (see Appendix). A satisfactory alternative is povidone-iodine (Betadine), which although similar in colour to iodine does not irritate these surfaces. Patients should be reassured that any staining of underclothes will not survive one wash.

Disinfection of apparatus, trolleys, tables, etc.—A sterile towel on a trolley can only act as a barrier to organisms as long as it remains dry; but liquids are so often spilt that it is advisable to use a trolley with a top of stainless steel or other impermeable material and to wipe it over with chlorhexidine/spirit immediately before the first towel is laid on it. Nylon film and autoclavable paper are now available and may be used instead of dressing towels. Before procedures requiring asepsis it is advisable to wipe over the surface of the X-ray table and the exposed parts of the X-ray screen or image intensifier with 0·5% chlorhexidine in spirit (70%). The polythene covers of sponge foam mattresses are similarly wiped over before use. This treatment is satisfactory for all contamination other than by tubercle bacilli or tetanus or anthrax spores. When a patient with obvious infection has been examined a similar cleansing routine is followed after the examination.

Disinfection of rooms.—Walls and floors are most conveniently disinfected by spraying with 1% 'Resiguard' until visibly wet and allowing 3–4 hours before re-occupation. In this situation 'Resiguard' has an advantage over stronger disinfectants, such as Hycolin, of not leaving a sticky surface when dry, hence washing is not obligatory.

Sterilisation of arterial catheters.—There are two levels of disinfection for polythene catheters; the ideal—achievable either by gamma irradiation or treatment with ethylene oxide—and the acceptable second best. The latter can be achieved by boiling but this is not practical because the polythene becomes softened and easily deformed. Since, however, catheter tubing is virtually sterile when manufactured and need not subsequently be heavily contaminated, acceptable disinfection is normally achieved by immersion for several hours in a mixture of domiphen bromide and chlorhexidine or slightly less predictably, in benzalkonium solutions (Roccal or Zephiran) or chlorhexidine-cetrimide mixtures (Cetavlon). Nevertheless it should be realised that these agents do not kill all organisms, particularly pseudomonads, and have virtually no effect on spores. Furthermore, unless the catheters are fed vertically into the storage fluid or flushed through with it the lumen of the catheter is not reached. It is therefore recommended that when catheters are prepared in batches before use, they should be handled under clean conditions by an operator who is scrubbed up. It is further suggested that before immersion in storage fluid the catheters should be flushed through with chlorhexidine/spirit and immersed in it for 1–2 minutes. We are well aware that even this practice does not provide full sterilisation but it is almost certainly an improvement on the methods at present widely used, which amount to no more than storing the catheters in a sterile liquid. If rapid disinfection is required the use of chlorhexidine/spirit alone as described above is adequate (immersing the catheter for at least 3 minutes) but in either case the spirit must be thoroughly flushed from the lumen of the catheter before use since it would otherwise promote clot formation. Permanent storage in chlorhexidine/spirit is not satisfactory as it tends to cause hardening and rigidity of the tubing.

Nylon tubing can be autoclaved; this is a considerable advantage but the existing range of tubing is unduly rigid. Both nylon and polythene catheters are discarded after use; the difficulty of cleaning and re-sterilisation is not worth the saving involved. Teflon catheters, which are now coming into use, are sterilised by washing through with hydrogen-peroxide and then autoclaving. They may be used repeatedly.

The polyvinyl connections used for arteriography provide another theoretical source of risk. They are inevitably filled with blood during arteriography and it is impossible to be certain afterwards that cleaning has removed all particles of the virus of serum hepatitis, should any be present. This risk is low in Britain because at present few members of the population carry the virus, but the risk is there and in some countries will be much higher. For this reason the polyvinyl should also be discarded after use.

Sterilisation of needles and instruments.—The method of choice for sterilising metal instruments is, as for sterilisation of syringes, by heating in a hot-air oven. Autoclaving can also be effective but it requires more care if full sterilisation is to be ensured. Boiling is much less effective since it will not kill most pathogenic spores. However, for many clinical purposes inactivation of all spores is not essential and boiling of metal instruments is then adequate. To ensure destruction of the virus of serum hepatitis it is necessary to clean and then boil for at least ten minutes any instrument which has been in contact with a patient's blood or serum.

Further reading

WILLIAMS, R. E. O., BLOWER, R., GARROD, L. P., and SHOOTER, T. A. (1960). *Hospital Infection*. Lloyd-Luke (Medical Books) Ltd., London.

APPENDIX

1. Povidone-iodine (Betadine—Berk Pharmaceuticals) is a chemical complex of iodine and polyvinylpyrrolidone which is effective against many spores as well as bacteria. It is available either as an antiseptic solution for skin disinfection, or, with added detergent, as a surgical scrub.

2. Chlorhexidine/spirit—is Chlorhexidine (Hibitane) 0·5% in spirit 70%.

3. Iodine—is either 1·5% Iodine in 80% isopropyl alcohol, or
   Iodine solution B.P. ($2\frac{1}{2}$% iodine in spirit).
   Isopropyl alcohol is more penetrating, more active when diluted with water and therefore more effective than ethyl alcohol.

4. Domiphen—chlorhexidine mixtures. Domiphen bromide (Bradosol) 0·5% (or benzalkonium 0·5%) with chlorohexidine (Hibitane) 0·1% in water makes a synergic mixture with is non-sensitising and is useful for disinfection of the vagina, glans penis and anus, for cleansing wounds or sinus mouths and for rinsing hands during 'sterile' procedures. Benzalkonium chloride (Roccal) can be used similarly and so can cetrimide and chlorhexidine 0·5% (Savlon).

5. *Roccal for catheter storage* (*also known as Zephiran*)
   Roccal   — 1 G.
   Tribasic sod. phos. — 10 G.
   Water   to 1 litre

6. *Hycolin*. This is a highly effective bactericidal (and often sporicidal) mixture of synthetic phenols. It is comparatively cheap, non-toxic and non-sensitising to the skin (but should not, of course, be used for skin disinfection).

CHAPTER 5

INTRAVENOUS UROGRAPHY[1]

THIS examination must be conducted by the radiologist if the maximum possible information is regularly to be obtained from it. Circumstances may make this difficult but it must never be thought that the unsupervised examination is a satisfactory alternative. The aim should be to make retrograde pyelography a comparatively rare investigation and this is seldom the case where intravenous urograms are not carefully supervised.

Modifying factors

1. *Combined liver and kidney failure.*—While renal failure is no longer a contra-indication to urography the additional presence of liver failure removes the usual route for excretion of contrast in azotaemia. In these circumstances urography is contra-indicated (Becker *et al.*, 1968).

2. *Previous reaction to contrast.*—This is seldom an absolute contra-indication; the matter is discussed more fully under 'Emergencies in the X-ray Department'.

3. *Myelomatosis.*—In the past myelomatosis has been thought to contra-indicate urography. It now seems that with present-day contrast media the risks are small (see Cwynarski and Saxton, 1969). Preparatory purgation and dehydration should be avoided since dehydration itself can produce renal failure in myelomatosis.

4. *Thyroid disease.*—Injection of contrast will vitiate subsequent iodine uptake studies or isotope therapy for 6–8 weeks.

5. *Pregnancy and the puerperium.*—An X-ray examination of the abdomen is undesirable during pregnancy. In addition to this it is difficult to obtain diagnostically helpful films because of the changes in the urinary tract which persist for some weeks after delivery. For this reason, whenever possible the examination should be deferred until at least 12 weeks after delivery and preferably for 16 weeks. When it is required a medium dose of contrast (see below) is given with a single erect film at 25–30 minutes after injection. The delay before taking the film will usually allow complete filling of the ureters and the erect position will minimise the distension of calyces and ureters

6. *Infancy.*—Rapid injection of contrast in infancy can produce a sharp rise in serum osmolarity (Standen *et al.*, 1965) and the injection should be given over at least 3 minutes.

[1] For a discussion of the theoretical aspects of urography and a full bibliography see Saxton (1969).

Preparation

1. *Bowels.*—Bowel preparation is discussed in Chapter I. Ambulant patients should be kept 'up and about' before the examination, and patients who are confined to bed should be made to sit up for as much as possible so as to reduce swallowed air in the bowel.
2. *Dehydration.*—Fluid restriction begins 24 hours before the examination—only three cups of fluid being allowed between then and 12 hours before examination; for the last 12 hours no fluid is allowed.
3. *Bladder.*—The patient empties the bladder before the examination starts.

For most cases this preparation should be adhered to but there are exceptions:—

1. In any examination undertaken at short notice, using medium or high doses, particularly in renal colic.
2. In renal failure it is *most important not to dehydrate the patient*. This also applies in myelomatosis.

Summary of Procedure

1. Plain film 17″ × 14″—enquire for history of allergy.
2. View plain film—inject contrast. Record the dose employed.
3. 5 minute film coned to kidneys, then apply compression or lower the head of the table.
4. 10 minute film coned to kidneys.
5. 15–20 minutes, release compression, then immediately take full length film.
6. Bladder film at 30–40 minutes if bladder not well filled on 15 minute film.
7. Film of bladder after micturition.
8. Special views as required. Delayed films in cases showing delayed excretion.

REMEMBER TO SHIELD THE GONADS WHEN POSSIBLE.

Preliminary film.—This may be a single 17″ × 14″ film or a 15″ × 12″ of the abdomen and 12″ × 10″ of the bladder. Usually a 17″ × 14″ is adequate, particularly if the exposure is made in inspiration for larger patients; but when ureteric calculi are suspected the pair of films diminishes the chance of stones being 'lost' over the sacrum.

The purposes of the preliminary film are:—

1. *To show the size, shape and position of the kidneys.*
2. *To show any opacities in the line of the urinary tract.* The normal plain film is taken in expiration; when opacities are visible over the kidney on the plain film a second film taken in inspiration may help to settle their relation to the kidney. An oblique film, however, also achieves this and usually with greater certainty.

3. *To show that the exposure is correct.* Unless the patient is young an incorrectly exposed film should be repeated. The lower range of kilovoltage is preferable (60–70 kV. for the average adult) since this gives better contrast and increases the chance of showing faintly opaque stones.

4. *To show ·that the colon and small bowel have been cleared of gas and faeces.* Occasionally the bowel is so heavily laden with gas or faeces that it may be thought advisable to defer the examination. This will only be necessary when tomographic facilities are not available.

5. *To demonstrate any other abnormalities.* It must not be forgotten that systems other than the urinary tract are also shown and the whole film—spine, pelvis, liver, spleen, lung bases, etc.—should be carefully scrutinised.

N.B.—Inquiries for a history of previous allergy should be made not later than the time when the preliminary film is taken. It must be made the responsibility of the radiographer or nurse either to ask the patient herself or remind the radiologist to do so.

Injection of contrast.—When the plain film has been checked and points of especial interest noted the injection may be made. Although for a normal patient who is well prepared, 20 ml. 45% Hypaque or 60% Urografin is often sufficient, there is, in practice, a strong case for the routine administration of larger doses. Many patients are incompletely dehydrated, or have obstructive disease or early renal failure, and their urinary concentration of contrast will be greatly improved by medium-level doses of contrast—e.g. 50 ml. of Conray 420 or Urovison for an average adult. This level of dosage is now recommended as the standard in this book and in some circumstances even larger doses are suggested (see Table 1). There is no advantage—except possibly convenience and lessened side-effects—in giving contrast as a drip, and as this is costlier, it is not recommended. Work now in progress suggests that the sodium containing media may achieve higher urinary concentration than comparable doses of methyl-glucamine bearing contrast. This favours such media as Conray 325 or 420, Hypaque 45 or Urovison which are wholly or mainly composed of sodium salts of contrast. When larger volumes of contrast are used their injection is facilitated by the use of wide-bore needles. In all cases it is an advantage if the dose employed is recorded for future reference.

As discussed earlier (Chapter 3) a small quantity ($\frac{1}{2}$–1 ml.) of contrast medium is injected initially, followed by the full injection a minute later if there is no reaction. It is helpful to talk to the patient during injection and take his mind off what he may be feeling. When injection is finished it is advisable, if time permits, to remain in the room (washing ones' hands, washing the syringe etc.), for a few minutes. When compression is used the apparatus may be fastened during this period.

The types of reaction to contrast medium and their treatment are discussed in Chapter 3.

Table I. Suggested doses of contrast based mainly on the use of Conray 420; comparable doses of other media may be calculated from Table VIII of the General Appendix. This table is reproduced from the British Journal of Radiology, by kind permission.

Indication	Suggested dose
Routine examination adult (70 kg)	50 ml. Conray 420
Routine examination elderly or obese	80 ml. Conray 420
Neonates and infants	3 ml./kg. of Hypaque 45— *slowly* or 2·5 ml./kg. of Conray 325— *slowly*
Children	20–40 ml. Conray 420 *slowly*
Renal colic	50–60 ml. Conray 420
To show ureters or in severe hydronephrosis	100 ml. Conray 420
Renal failure	100 ml. Conray 420 (or 150 ml. Conray 280) +tomography. NO DEHYDRATION
Haematuria, renal masses, suspected polycystic disease, non-visualised kidney, renal trauma	100 ml. Conray 420 with immediate tomography. No compression

Compression.—Efficient compression is of value in obtaining good filling of the calyces and subsequently, on its release, of the ureters. There are a number of objections to its use:—

1. *Discomfort for the patient.* Some patients, notably those with renal infections or scars of recent operations, cannot tolerate compression and high dosage is then advisable. Most patients dislike it but can be persuaded to put up with it. Since it may add to the quality of the examination and in some cases may save a retrograde pyelogram, it is worth some discomfort.
2. *It may give a false impression of hydronephrosis.* This only holds if technique is unsatisfactory. A 5 minute film should be taken before compression is applied and will, except where there is delayed excretion, show the true state of the calyces. (This delay before applying compression also allows contrast to displace urine down the ureters and avoids dilution of contrast.) At least one film is taken after release of compression and this also will usually show the calyces to be of normal size; if not a full-length film after micturition should do so. In practice, the calyces are seldom more than well filled by compression and it is unusual for there to be any doubt.

3. *Impaired venous return may cause 'faints'.* This is rare and is treated by release of compression and lowering the head of the table.
4. *It is not suitable for use by radiographers on their own.* This objection is largely valid. Compression with a Bucky band and pads is permissible for radiographers but is less efficient than the balloon type of compression. Few radiographers seem capable of applying the latter effectively and since it may cause fainting a radiologist should in any case be at hand.

The decision to use compression is one which each department makes for itself. If it is not used the bowel preparation must be as good as possible and it is essential to give medium doses of contrast as a routine. In addition the head of the couch may be lowered, after the 5 minute film, into the steepest possible Trendelenburg position compatible with complete safety.

Our own preference is for using compression but in advocating it we make certain conditions:—

1. It should be of the type which encircles the patient (see Appendix and Fig. 5) enabling him to be turned into the oblique position while the compression is still applied. A Bucky band and pads do not allow this and in addition cannot be applied with a known pressure.
2. An inflatable rubber bladder is used and should be large, extending all the way across the abdominal plate. Small pairs of bladders are unsatisfactory because they will give a high pressure reading with a small amount of air, i.e. as soon as their own walls begin to be stretched. Moreover they are difficult to apply effectively—being small, pressure is only exerted over a narrow area.
3. Compression is applied after the 5 minute film has been taken. It is normally only kept on until just before the 15 minute film is exposed. It may be maintained after this if a second injection has been given to improve the contrast. It is pointless to keep it on for longer when there is hydronephrosis since the pelvi-ureteric system is already obstructed.
4. The pressure in the balloon need not be high *if the belt has been firmly applied.* 60–70 mm. Hg. is sufficient in the slender but higher pressures are needed in fat patients, up to about 120 mm. Hg.

Timing of films.—A useful average series consists of coned kidney films at 5 and 10 minutes in expiration with a full length film *immediately* compression has been released (or the table brought level from the Trendelenburg position when no compression has been used). The timing of the full-length film depends on the dosage employed since larger doses produce greater urine flow and so more rapid bladder filling. When the higher levels suggested earlier are given adequate bladder filling is usually obtained by 15 minutes after injection and the full-length film can be taken then; this is particularly useful in achieving a rapid turnover of patients. With lower doses it is advisable to defer this film to 20 or 25 minutes. The 5 and 10 minute films are viewed as soon as possible so that:

(i) Any corrections in centring or exposure can be made.
(ii) Extra views may be taken where indicated.
(iii) Should concentration or calyceal filling be poor a second 'booster dose' of 50 ml. of contrast may be given to cases who have only received an initial 50 ml.

Centring of coned films is made easier if the position of the film is adjusted in relation to the iliac crests. In most cases a $12'' \times 10''$ film crosswise with its lower edge at iliac crest level will include the kidneys— the X-ray tube is then centred to the film. Abnormalities in position of the kidneys should be noted on the plain film and the coned film position adjusted accordingly. The centring for films taken in inspiration is $1-1\frac{1}{2}$ ins. (3–4 cm.) lower to allow for descent of the kidney.

In the average patient this series will be sufficient provided that renal function is normal. Impaired concentration is an indication for higher doses and further films with tomography as discussed later in the chapter. When the full-length film is taken relatively early the bladder may not be completely demonstrated because it is not filled out adequately. When this happens a further film coned to the bladder should be taken about 20–30 minutes later; the patient may be allowed to drink to increase the urinary flow. This manoeuvre may also be useful in cases where there is an apparent extrinsic impression and it is desired to show that the bladder is in fact, capable of normal distension.

In the presence of a bladder residue of urine before the contrast has been injected the contrast medium tends to collect in the posterior part of the bladder and the anterior part is not outlined. In this case the patient is kept on the table for a further 30 minutes and is then turned prone and after this encouraged to walk about. This manoeuvre will usually produce adequate outlining of the bladder but with a really large residue cystography by catheter may be required for satisfactory demonstration of the bladder.

With younger patients and particularly for 'exclusory' examinations the numbers of films taken can be much reduced. A coned 10 minute kidney film and a full-length 15 minute film are usually adequate.

Poor concentration and delayed excretion.—It is important to distinguish between poor concentration due to failing kidneys or inadequate dehydration and delayed excretion due to obstructive lesions. In the former the pelvi-calyceal patterns appear quite early i.e. by 5 or 10 minutes and show no dilation. The bladder opacifies early and fills out with a quantity of dilute contrast. In these cases there is no object in taking delayed films unless a second dose of contrast has been given in the hope of improving the concentration. In obstructive lesions there is a true delay in excretion. At 15 minutes only the calyces may have filled and the renal pelvis may remain empty for 1–2 hours or more. If the abnormality is bilateral the bladder does not opacify until the contrast has run down the ureters and this may not be for some time. In such cases it may be necessary to take films at 45, 60, 90 or 120 minutes. Some excretion is often visible on the later films when early films show no contrast or only a nephrographic 'blush'. A second injection of contrast and films up to 6 or more hours later may then produce quite useful delineation. For practical purposes however if no evidence of *any* excretion is seen at 2 hours the kidney is regarded as non-functioning. The management of cases with hydro-nephrosis is discussed further below.

Techniques for special problems

Extra plain films have already been mentioned. Abnormalities seen during the course of the examination may indicate the need for special views so that the 5 and 10 minute films are viewed as soon as possible. Of course, the plain film findings—e.g. probable calyceal stone—may indicate the need for an extra film, i.e. an oblique view, without first seeing the 5 and 10 minute films.

A. Oblique views of kidneys. (30° left side up for right kidney and vice-versa.)

These may be of value in:—

1. Possible calyceal irregularities or deformities.
2. Localising a calyceal stone as in an anterior or posterior calyx.
3. Demonstrating the pelvi-ureteric junction.
4. Possible space-occupying lesions or irregularities in the renal outline, e.g. cortical scar.
5. Opacities whose relationship to the kidneys is uncertain.
6. Cases where gas in the bowel obscures calyces.

It is not usually advisable to cone down on an individual kidney for this view. A 12″ × 10″ cassette placed crosswise will give, if covered, a useful view of the opposite kidney as well as the one under particular scrutiny.

B. Films of the whole abdomen.—If there is any reason to suspect an ectopic, particularly a pelvic kidney, films of the whole abdomen are taken instead of coned kidney films. Similarly, when seeking to show the ureter in relation to a suspected calculus it is advisable to take full-length films throughout as ureteric filling may only occur on one film.

C. The management of cases with hydronephrosis.—It is not sufficient to demonstrate the presence of hydronephrosis; it is imperative to determine the site of any obstructing lesion and, where possible, its nature. This may involve taking films for a considerable period and possibly a prone film as discussed below. The use of medium or high contrast dosage is essential if optimal visualisation is to be achieved.

Prone films.—In the presence of hydronephrosis the contrast, being of high specific gravity, pools in the most dependent parts of the calyces and does not, because of the stasis, move forward. In such cases one should wait (if necessary giving a second injection of contrast) until films at 45 minutes, 60 minutes, 90 minutes or 2 hours show that the contrast has 'built-up' in the calyces and pelvis. Then the patient is turned prone, told to take deep breaths and a film is taken after a pause of $1\frac{1}{2}$–2 minutes— varying with the interval after injection at which the film is being taken. This will bring contrast to the pelvi-ureteric junction and if the block is in this region it will be demonstrated. The patient is turned supine after this film. Sometimes it is evident that the obstruction is lower, in the ureter, and that the contrast has not reached it. If so the patient should sit up and lean forward for about 3–5 minutes—again according to how late the film is. This is followed by a supine film. Prone films may also be useful for outlining grossly dilated ureters.

If contrast medium is not allowed time to accumulate in the calyces and pelvis the prone film is less useful because the small amount of contrast present is diluted by the static urine in the hydronephrosis.

D. The detection of early obstruction.—Routine use of medium doses helps to reveal cases of minimal obstruction because such obstruction is more obvious under conditions of high urine flow. At times, however, there may still be doubt, particularly as to the presence of pelvi-ureteric obstruction, and two techniques are then available.

Erect films.—In most patients the kidneys drain completely in the upright position and if an erect film shows adequate emptying significant obstruction is unlikely. When there is still a residue in the renal pelvis on the erect film or when the patient gives a history of pain after high-fluid intake diuretic urography is undertaken.

Diuretic urography.—In this technique further contrast (e.g. 50 ml. Conray 420) is given and a diuretic such as frusemide (Lasix) 20 mg. is injected intravenously. Films are taken every 5 minutes until contrast has been washed out of the kidneys. Cases with incipient obstruction often show increased pelvi-calyceal distension, delayed washout of contrast and will often complain of loin pain.

E. Tomography of the kidneys.—This is of value:—

1. When gas obscures the kidneys.
2. When the renal outlines are not clear and their delineation is important, e.g. in hypertension or with possible renal masses.
3. When contrast is faint, particularly in renal failure.

In thin adults lying supine, cuts at 7–9 cm. from the table top usually demonstrate the kidneys; for more average adults 9–11 cm. are suitable and in fat patients 11–13 cm. or more. Further cuts forward or backward are added if required. In cases where the kidneys are hard to locate it is helpful to bear in mind that they usually lie at the level of the posterior bodies and pedicles of L.1, 2 and 3 though they may be more anterior than this in heavily built lordotic patients. Pelvic kidneys and transplanted kidneys are commonly level with the superior pubic rami.

When tomography is used as a routine, which is invaluable when the examination is unsupervised by radiologists, it is carried out at about 12 minutes after injection.

Nephrotomography

Lesions of the renal parenchyma are often well demonstrated by giving a high dose of contrast followed by immediate tomography. It is advisable to take at least 5 cuts so as to show the entire kidney at the first attempt. The exposure should not be too 'soft' or detail is inadequate. This technique can be used *de novo* or to supplement an examination already in progress. It is particularly useful for suspected renal masses, in polycystic disease or in demonstrating that a non-visualised kidney still shows some evidence of excretion into a thin rim of distended parenchyma.

Zonography

In this technique a tomographic exposure with a reduced angle of swing is employed (5°–15°). This demonstrates a greater thickness of tissue and allows some shortening of the exposure. A single film will often show the kidneys in their entirety. These advantages make the method particularly suitable for children.

F. Oblique bladder films.—(Patient rotated about 40°, tube angled 20–25° towards the feet.)

These are taken:—

1. To show the size and extent of diverticula of the bladder.
2. To confirm or exclude possible bladder filling defects seen on the A.P. film, including prostatic impressions.
3. In suspected ureterocoele (with the same side raised).
4. *To determine whether opacities are in the ureter or not.*

When there is a possible ureteric stone in the pelvis on the plain film and *when excretion is normal on that side* (as judged from the 5 minute film) one must be alert to 'catch' the ureter when it is full. This means taking an A.P. full length and an oblique film of the pelvis (i.e. with the side opposite the opacity raised) *immediately* compression is released or at 10 minutes after injection when compression is not used. In the latter case an extra manoeuvre which may help to show the ureter is to tell the patient to prop himself up on his elbows, to take two or three deep breaths and then lie back for the full length film.

Later in the examination, if the bladder is full, it may obscure opacities low in the line of the ureter. In this case the opposite oblique—i.e. with the same side raised—is then better. When there is still doubt an after-micturition film with or without an oblique view may be helpful.

When excretion is delayed on the side of the opacity the problem is to achieve ureteric filling down to the region of the opacity. This is discussed above under the management of cases with hydronephrosis and below under 'Emergency urography'.

Angulation of the tube towards the feet is important since otherwise the pubic rami overlie the lower ureters or bladder edge.

G. Excretion urography in infants and children.—*Preparation.*—This should not be drastic. In healthy infants the interval between feeds is adequate and in neonates or sick infants no attempt is made to dehydrate. In older children eight hours dehydration is sufficient; when possible they should be kept up and about to minimise swallowed gas in the intestinal tract.

Sedation is usually needed for infants and for most small children under 3 or 4 years. Examination of a restless child may cause wastage of film and fruitless irradiation. Sedation may not be necessary with radiographers who are expert in handling children.

Children over 7 or 8 may be prepared in a similar way to adults but an aperient and limitation of fluids for 12 hours are usually all that is required.

To reduce the obscuring effect of intestinal gas it has been found of value to distend the stomach with gas. In bottle-fed infants this is usually achieved by giving them a feed in the supine position immediately before and after injection. The swallowed air will distend the stomach—especially if the bottle is intermittently tilted the wrong way so that the child sucks air. Giving the feed in this way also reduces the danger of excessive dehydration.

An alternative manoeuvre which is applicable to older children is to give a carbonated drink immediately injection of contrast is finished. This technique is not always successful when carbonated fluids are used. It is probably better to arrange prior intubation of the stomach; air can then be injected just before the 5 minute film. The upper age limit for this manoeuvre is about 8 years; after this the stomach distends less fully and does not overlie the kidneys so completely. An oblique film (left side raised) is taken at 5 minutes as well as the A.P. since the right kidney is not always covered by the gastric 'window' on the straight film. (Hope *et al.*, 1957.)

Tomography or zonography provide alternative methods of showing kidneys obscured by gas. The child must be still and heavy sedation is therefore needed. In small infants cuts at 3–5 cm. from the table top are used; older children require cuts further forward according to their size. Using zonography a single cut taken at 3 minutes is often sufficient.

Injection of contrast. Where possible this should be intravenous,[1] in the dosage suggested in Table I.

Coned kidney films are taken at 5 and 10 minutes and later if necessary. A full length film is taken at 15 minutes.

H. Emergency urography.—This is indicated:—

1. For renal colic when there is no visible calculus on the straight film.

---

[1] Subcutaneous or intramuscular injection are alternatives. Subcutaneous injection is achieved by diluting 30 ml. of 45% Hypaque (or equivalent) with 70 ml. of 5% dextrose and adding an ampoule of 1000 units of hyaluronidase. 15 ml. are injected widely in each subscapular region with massage of the injection sites. For intramuscular injection Fournier (1966) places the infant in the prone position and introduces a needle into each buttock. Half an ampoule of suitably diluted hyaluronidase is then injected through each needle. One minute later the undiluted contrast, to which a further ampoule of hyaluronidase has been added, is injected, half into each needle; the injection sites are then massaged.

2. When there is doubt as to whether a pain is renal or not and the plain film is uninformative.

Technique.—No special preparation is used nor any compression applied. 50–60 ml. of Conray 420 or equivalent are injected and a full length film taken 15–20 minutes later. Should this film show normal excretion on both sides no further films are needed and it can be taken that any pain felt is not due to obstructive renal colic. If the examination is not conducted during an attack of pain it becomes more difficult to evaluate a 'normal' result; all that may be evident between attacks is a slight overfilling of the ureter on the affected side; the return to normal appearances may take as little as an hour.

Delayed excretion indicates that there is probably ureteric obstruction and further films may be taken to show the site of the obstruction. The timing of such films depends on the delay in excretion. Thus if at 20 minutes contrast has reached the renal pelvis but not the ureter a film may be taken at 45–60 minutes. If the latter shows only partial filling of the ureter a further film is taken after the patient has sat up for 2–3 minutes. If at 25 minutes contrast is only in the calyces one should wait 60 minutes before taking the next film. If there is only a nephrographic blush it may be necessary to continue taking films up to 3, 6 or 12 or even 24 hours before there is filling down to the point of ureteric obstruction. When there is this sort of delay a further injection of 50 ml. contrast at 6 hours is recommended. The bladder should be emptied before taking delayed films; this prevents the masking of the lower ureters by the distended bladder. As soon as the point of ureteric obstruction has been shown the examination is terminated.

I. Urography in hypertension.—Two points require attention in examining cases of hypertension. The first is the importance of demonstrating the parenchyma adequately; if necessary tomography must be used. The second is the desirability of detecting cases of significant renal artery narrowing, where possible. The modifications of technique which may be used are (i) early films (rapid sequence technique), (ii) washout urography.

(i) *Early films*

Films of the kidneys are exposed 1 and 2 minutes after injection of contrast. This may provide evidence of asymmetrical excretion of contrast.

(ii) *Washout urography*

The kidney distal to a renal artery narrowing may show altered function with diminished urine flow and hyperconcentration of contrast. The difference between the normal and the abnormal side can be accentuated by producing a diuresis and is probably maximal after a urea-saline load (Amplatz, 1964). A slow-running intravenous drip of 40 g. urea (Ureaphil-Abbott) in 500 ml. of saline is set up. 50 ml. of Conray 420 (or equivalent) is injected rapidly. If necessary a series including rapid sequence films is exposed and the presence of normal renal function and anatomy is confirmed. (This may not be necessary if conventional urography with early films has already been undertaken). The urea-saline drip is then allowed to run in as fast as possible and films of the kidneys are exposed every 4 minutes until one or both kidneys no longer shows any appreciable contrast.

J. Urography in renal failure.—Renal failure is now accepted not as a contra-indication to but as an indication for urography provided that high dosage and tomography are employed and it is appreciated that certain points are important:—

1. *Dehydration.* This *may be dangerous and must not be used* in the patient with a raised blood urea: the result can be acute renal failure. The difficulty here is to avoid the automatic application of preliminary dehydration by nursing staff accustomed to its use in normal subjects. The preparation slip must therefore carry a warning to this effect (see below).

2. *Dosage.* High dosage of contrast should be used, i.e. 100 ml. Conray 420 or an equivalent dose of other media. In patients who have hypertension or heart failure the methylglucamine salts of contrast (Conray 280 or Urografin 60% or 76%) are preferable to avoid an excessive sodium load.

3. *Timing of films.* There is no particular urgency to obtain tomographs in cases of chronic renal failure but in most cases satisfactory visualisation is obtained within a few minutes of injection. In obstruction, as would be expected, delayed films may be needed to demonstrate the pelvi-calyceal system fully. In acute renal failure, particularly oliguric renal failure, there may be considerable value in delayed films, up to 24 hours or later. Not only do some cases show clearer outlines of the collecting systems but the pattern of change of the nephrogram may give a valuable indication of the nature of the disease process in the kidney.

4. *Prior use of dialysis.* There is now some evidence that better visualisation is achieved after dialysis than before (Matalon and Eisinger, 1970). When possible the examination should be deferred until effective dialysis has been carried out.

Difficulties.—The main difficulties which arise in conducting an I.V.U. have already been mentioned—gas in the bowel, poor dehydration etc. Pelvi-calyceal 'spasm' is sometimes a problem. It is less frequent when compression is used and it is also less common when medium or large doses of contrast are given.

One other problem which occasionally arises is that of *delayed excretion due to a fall in blood pressure*; this may result from a reaction to contrast or from a faint due to the venepuncture. If the blood-pressure falls to the region of 70 mm. Hg. systolic or lower, excretion of contrast stops. It is therefore necessary to wait until the blood-pressure is restored to normal before a urogram can be obtained.

References

AMPLATZ, K. (1964). *Radiology*, **83**, 816.
BECKER, J. A., GREGOIRE, A., BERDON, W. and SCHWARTZ, D. (1968). *Radiology*, **92**, 243.
CWYNARSKI, M. and SAXTON, H. M. (1969). *Br. med. J.* **1**, 486.

FOURNIER, A.-M. (1966). *J. Radiol. Electrol.* **47**, 507.
HOPE, J. W., O'HARA, A. E., TRISTAN, T. A., and LYON, J. A. (1957). *Med. Radiography and Photography*, **33**, 48.
MATALON, R. and EISINGER, R. P. (1970). *N. Engl. J. Med.* **282**, 835.
SAXTON, H. M. (1969). *Br. J. Radiol.* **42**, 321.
STANDEN, J. R., NOGRADY, M. B., DUNBAR, J. S. and GOLDBLOOM, R. B. (1965). *Am. J. Roentg.* **93**, 473.

APPENDIX

Instructions to patient or ward

24-*hour preparation for Urography* (*kidney examination*)

Your X-ray appointment time is .... on.....................
Eat normally (no soup) but drink ONLY THREE cups of fluid (tea, milk, soup, beer, etc.) in the 12 hours from .... a.m. on ....
until .... p.m. on.....................
After that eat ONLY DRY FOOD and drink NOTHING AT ALL until coming to the X-ray Department.

N.B. TO WARD STAFF: PLEASE *DO NOT* DEHYDRATE PATIENTS WHO HAVE RENAL FAILURE OR POSSIBLE MULTIPLE MYELOMATOSIS.

Take the tablets at the times indicated. They will cause your bowels to act which is most important for the success of the examination.

The examination will take 1½–2 hours.

Bowel preparation

This is discussed in Chapter I.

Compression (*M.D.*)

The essentials of a compression system are shown in Fig. 5. They comprise:—

1. A perspex or metal plate to cover the pelvis. (In the model devised by Prof. J. H. Middlemiss lead protection is incorporated in the plate so that in female patients protection cannot be forgotten).

2. A broad rectangle of canvas which goes under the sacrum and from which broad straps encircle the abdomen.

3. A large rubber bladder which lies under the plate, a manometer to register the pressure, and hand bellows for inflation.

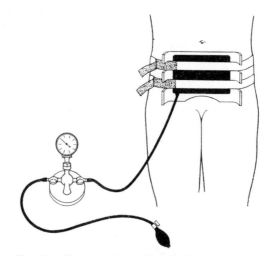

FIG. 5.—Compression apparatus for intravenous urography. Note that the lower edge of the plate lies just above the pubis.

The lower edge of the plate normally lies just above the pubis. The straps are tightened firmly before inflating the balloon; in thin patients a little padding may be needed between the straps and bony prominences.

Tray or trolley setting

For urography a tray serves as well as a trolley.

Sterile tray

Gallipot—Chlorhexidine/spirit.

3–4 cotton wool swabs.

Kidney dish—50 ml. syringe with an eccentric nozzle.

Drawing-up cannula. Needles, hypodermic size Nos. 1 and wide bore e.g. Gillette 19G—1 R or S—TW.

Unsterile

Contrast, ampoules or vials in a bowl or glass of hot water.

Ampoule files.

EMERGENCY DRUGS.

CHAPTER 6

RETROGRADE PYELOGRAPHY

THE use of retrograde pyelography as a general survey examination of
the renal tracts is most undesirable and with the improved media now
available for intravenous urography it can seldom be justified. It is
properly undertaken to elucidate a specific problem raised by the urogram.
It is essential that this problem should be fully understood before the
examination is begun so that the investigation may be planned to yield the
maximum information.

Retrograde catheterisation.—When the only relevant question is whether
or not an opacity is in the ureter, an A.P. and oblique film (raising the side
opposite the catheter) will be sufficient. In such cases contrast should not
be injected unless there is some other indication, for this always increases
the risk of introducing infection.

Summary of retrograde pyelography

1. Plain film to show catheters.
2. Pull catheters down if necessary.
3. Gently aspirate urine if possible.
4. Inject contrast deliberately—average pelvi-calyceal system requires
   4–7 ml.—expose films with injection almost arrested. Always stop
   injection if the patient complains of pain.
5. View films, repeat with oblique views if necessary.

REMEMBER TO SHIELD THE GONADS WHEN POSSIBLE.

Retrograde pyelography.—The patient usually arrives in the X-ray
department from the theatre on a trolley, although in some hospitals
radiography is carried out in a cystoscopic room. A preliminary film is
taken to show the catheter position. While waiting for this to be developed
the urogram is reviewed, and, if the patient is to be screened, the time is
used for dark adapting. (An image intensifier removes the need for dark
adaptation and makes it possible to adjust the catheter under screen
control. It may, however, still be advisable to take a control overcouch
film to show the relationship of possible calculi to the catheter.)

Screening vs. Overcouch tube for obtaining films.—Each method has
advantages. Screening during injection offers the following benefits:—

1. Certainty of getting adequate filling without overfilling or backflow.
2. Quickness and accuracy in the taking of oblique views.
3. Better ureteric visualisation because multiple spot films can be taken.
4. Good control when the catheter is placed low in the ureter.

Overcouch films on the other hand have the advantage that:—

1. Sharper, 'brighter' films are obtained.
2. Time is not spent in dark adapting in departments without image intensification.

Most retrograde pyelography is performed 'blind' but having used both methods we have no doubt that screening is the better. Where image intensification is available screening is unquestionably superior, for it has the added advantage that the surgeon may watch if he so desires and that pelvi-calycine and ureteric movement can be observed. It is, of course, possible to commence the examination by screening and end by taking an overcouch film.

Adjusting the catheter.—After inspecting the plain film the catheter is withdrawn, if necessary, in order to bring the tip just below the pelvi-ureteric junction. It may be difficult to judge exactly how much to pull down if the catheter is coiled in the bladder; an attempt should be made to assess the length of the loop in the bladder and the excess in the kidney. When the catheter tip is thought to be in the kidney substance it is especially important to bring it down sufficiently and a second plain film may be needed. 'Knots' in the catheter are almost always apparent and pull out without trouble but in practice if the *tip* is placed satisfactorily it is usually better to inject before pulling the catheter out, otherwise there is a risk that the tip may pull out before the loop of the 'knot'.

Injection of contrast.—The handling of the catheter must be scrupulously aseptic since the introduction of organisms is most undesirable and in hydronephrosis may destroy the affected kidney (Hanley, 1960). The catheter ends are in test tubes strapped to the inner aspect of the thighs— the Red tipped catheter should be that in the Right ureter when there are two.

The operator, wearing a lead apron and a mask scrubs up and drapes a sterile towel across the thighs. The catheter just above the test-tube is grasped with a swab soaked in chlorhexidine-spirit and is lifted out of the tube; another spirit swab is used to wipe it to its tip and is laid on the towel, being held down with a towel clip or spirit swab.

The needle used to introduce contrast into the catheter should fit 'snugly'. Tapering pyelography cannulae are best but ordinary needles are adequate if the catheter grips them firmly. The needle should be attached to the syringe *before* introduction into the catheter as it is very difficult to avoid including an air bubble when syringe and needle are being fitted together. Care is needed to remove all bubbles from the syringe. When bubbles adhere to the wall of the syringe, draw air into the syringe and rock it so as to bring the bubbles off, then express all the air. As a further precaution use a syringe with an eccentric nozzle, keep the nozzle well down during injection and do not inject the last $\frac{1}{2}$–1 ml. When the needle is being fitted into the catheter, gentle injection of a few drops of contrast may flush out any air at the catheter mouth. The quality of films may be improved by deliberately aspirating urine from the renal pelvis before injecting contrast. This is also helpful in that it can give an indication of the capacity of the renal pelvis and hence of how much contrast may be injected in safety. The method does, however, increase the risk of introducing air bubbles into the system.

Contrast medium.—It is commonly recommended that contrast be dilute (e.g. 30% Urografin, 25% Hypaque) in order to avoid 'blotting out' small filling defects. This is valid in cases where the catheter is well placed and the pelvi-calycine system is near to normal in size. There are, however, some indications for a denser contrast, (60% or 76% Urografin, 45% or 65% Hypaque[1]):—

1. *In hydronephrosis.*—Here the contrast is diluted by the static urine and therefore dense contrast is best. Moreover the important region is the narrowed area causing obstruction and in such an area dilute contrast will not give clear definition. A further argument for the more concentrated medium is that a smaller volume is used—an important point in hydronephrosis.

2. *In ureteric problems.*—The normal ureter is a narrow structure and may not show clearly if dilute contrast medium is employed. Higher concentrations (e.g. 60% Urografin, 45% Hypaque) provide better delineation and unless the ureter is considerably dilated there is no risk of concealing filling defects—in fact, their definition may be improved.

3. *When the catheter is low in position.*—In this case one can only hope to fill the pelvis and calyces by forceful injection of contrast up the ureter so that it mixes with the urine in the pelvis and calyces. With the head tilted down it is usually possible to obtain adequate filling in this way. 45% Hypaque is the most satisfactory for this since it is less viscous and will therefore inject and mix more easily.

Technique of pyelography

Little need be said concerning screening technique except to point out that the commonest fault of the inexperienced is to expose the films before the renal pelvis and upper ureter are adequately filled. A useful general examination is provided by two straight exposures on a 'split' 12″ × 10″ film followed by two oblique exposures—one oblique each way—on a second 'split' 12″ × 10″ film (if the lower ureter is also to be examined as well a 14″ × 14″ film is used). When two catheters are present, i.e. one on each side, it is advisable not to inject more than 4 ml. at first, for it is possible to watch the wrong side for contrast medium while the opposite side overfills!

In 'blind' pyelography one first assesses the size of the pelvi-calycine system from the urogram. A normal system with a well sited catheter will

---

[1] Sodium iodide is still widely used but is more irritating than the organic media. Moreover, if significant extravasation occurs, serious injury may result.

take about 4–7 ml. This refers to the amount reaching the kidney. Allowance must be made for leaky needles, syringes etc. and a larger volume will be injected if the pelvi-calcine system is larger than average. Injection is made deliberately but not too slowly taking about 15 seconds for 5 ml. and only slowing or stopping if the patient complains of pain. When about 3 ml. are left before the planned amount has been injected, call out 'Breathe in—breathe out, stop breathing, *quite* still' and then to the radiographer 'Take'. Injection is continued while instructing the patient but is slowed very much for the actual exposure and stopped immediately after the film has been exposed. With a good catheter position the patient should at most feel slight discomfort towards the end of injection; any appreciable pain is the signal to slow or stop and take a film.

When the catheter is low in position a greater volume and more forceful injection are needed, using 45% Hypaque. The patient's head is lowered by tilting the table. With a catheter tip over the sacrum inject 15–20 ml. as rapidly as the patient can endure before exposure. For intermediate positions proportionally less is used.

It may be evident from the urograms that oblique views of the kidney or pelvi-ureteric region will be needed and these may then be taken straight away, if all has gone well with the A.P. film. A further 2–3 ml. is adequate when the side under examination is lowered but when the patient is rotated to the opposite side the pelvis tends to drain and about 4 ml. are injected.

In either technique when suitable films have been taken the syringe and needle are removed, the catheter is 'anchored' with a towel clip or wet swab and the end left to drain on to gauze swabs. A sterile towel is laid over it.

The films, when ready, are carefully studied. Radiographic and technical adequacy are important but the films are only truly satisfactory if they answer the question which the examination set out to answer. If they do not it may be due to wrong exposure or underfilling or it may be that oblique or lateral views or even prone films are needed. Only when all that is necessary has been achieved is the patient allowed to leave the department. Before removing the catheter the radiologist should make sure that it will not be further needed, as for example in obtaining divided urine samples for culture.

Special problems

Hydronephrosis.—The dangers of retrograde pyelography in hydronephrosis are well recognised. The sudden extra distension of an already distended system may precipitate complete obstruction with rapidly supervening infection, (Hanley, 1960). For this reason the catheter tip should always be withdrawn below the pelvi-ureteric junction before any contrast is injected (Anderson, 1962). In fact, with good urographic technique the presence of hydronephrosis and the thickness of kidney substance can nearly always be demonstrated and then the only remaining problem may be the site and nature of the obstructing lesion. If this can be shown with a small volume of contrast medium it is unnecessary to fill the whole pelvi-calycine system. Oblique or lateral films are helpful in showing the narrowed area (Rolleston and Reay, 1957). As already noted the most concentrated contrast which can be injected is used. An alternative technique which has much to recommend it is the use of the bulb catheter (see below).

It may in fact be possible to demonstrate the pelvis and calyces completely even when only a small amount of contrast medium has passed through the narrowed area. This is achieved by rolling the patient prone but because it is difficult to perform this manoeuvre without making the catheter unsterile it is left until the end of the examination.

In cases of doubtful hydronephrosis or after plastic repair a film is taken ten minutes after removal of the catheter and when the patient has been sitting up. This will show how well the pelvis is draining; normally the pelvis is virtually empty under these conditions (Anderson, 1962).

Ureteric demonstration.—Sometimes the problem is to delineate a ureteric abnormality. This is usually possible if the film is taken during injection but dense contrast and multiple spot films taken with screen control make adequate pictures more certain. When the catheter has passed through a narrowed area it may be difficult to show the lower end. Once the upper part has been demonstrated one can usually show the lesion and its lower margin by withdrawing the catheter, injecting as it is withdrawn and taking films during injection. Even if the narrowed area itself cannot be demonstrated its lower limit must be shown. Here too the bulb catheter is of value.

Alternative method

A bulb catheter inserted into the ureteric orifice with retrograde filling of ureter and pelvi-calyceal system has several advantages as a routine. It is less traumatic; the ureter is better displayed and the risk of precipitating obstruction in patients with pelvi-ureteric narrowing is reduced. Except under unusual circumstances, screening with spot films becomes impossible and the radiologist can only contribute in an advisory capacity. Where such an examination is inconclusive the ureter is catheterised and the examination conducted along the lines set out earlier.

Nephrostomy pyelogram.—Here the contrast is introduced into a tube which is connected to a drainage bag. A technique similar to that employed for T-tube cholangiography is used. The tube is clamped at a convenient site. An assembly consisting of an intramuscular needle, an arteriographic connection and a syringe full of contrast is prepared and the needle is inserted into the drainage tube above the clamp. Filling of the pelvi-calyceal system is controlled fluoroscopically.

Difficulties and complications.—The principal difficulties of the examination are those inherent in the surgical aspects, i.e. finding the ureteric orifice and passing the catheter a suitable distance up the ureter. The catheter may penetrate the renal substance partially or completely; the ureter may also be penetrated.

These complications are reflected in the radiological aspects of the examination. A traumatic catheterisation may cause oedema or spasm of the ureter or a calyx. Blood clot following penetration of the kidney can give rise to a filling defect in the renal pelvis. Injection of contrast with the catheter tip in an unsuitable position will produce extravasation. When spasm of pelvi-calycine or ureteric muscle is suspected 0·02 G. of hyoscine N-butylbromide (Buscopan) may be injected intravenously and the examination repeated in 5 minutes time.

Even when the catheter is satisfactorily positioned a forceful injection or injection of too large a volume can produce 'back-flow' into the kidney. This, whether pyelo-sinous, pyelo-venous, pyelo-interstitial or pyelo-lymphatic, seldom appears to cause serious damage to the kidney but it sometimes interferes with satisfactory delineation and renders the examination valueless.

The danger of introducing infection has already been mentioned and the risks of the examination in cases of hydronephrosis have been pointed out. One final point may be made; observations by Hope and Michie (1959) suggest that following retrograde pyelography in children a temporary ureteric obstruction and hydronephrosis may occur in some cases. This serves again to emphasise that retrograde pyelography is an examination which should only be undertaken for definite indications and that every effort should be made to achieve a satisfactory and conclusive examination by intravenous urography.

References

ANDERSON, J. C. (1962). *Proc. R. Soc. Med.*, **55**, 93.
HANLEY, H. (1960). *Lancet*, ii, 664.
HOPE, J. W., and MICHIE, A. J. (1959). *Radiology*, **72**, 844.
ROLLESTON, G. L., and REAY, E. R. (1957). *Br. J. Radiol.*, **30**, 617.

APPENDIX

Sterile trolley

2 towels.
Gauze and wool swabs.
1 gallipot with chlorhexidine/spirit.
1 receiver with 2 × 20 ml. syringe and drawing-up cannula.
Selection of needles, No's. 1, 2, 12, 14, 16, or
Selection of pyelography cannulae (G.U.).
Paper towel for drying hands. Gloves.

Lower shelf

In a bowl ⌠ 6 × 10 ml. ampoules of Urografin 30% or Hypaque 25%
of hot ⟨ 2 × 20 ml.   ,,   ,,   ,,   76% or Hypaque 65%
water ⌊ 2 × 20 ml.   ,,   ,, Hypaque 45%
Ampoule files.
Torch (with red filter).
Buscopan 2 ampoules each containing 0·02 G. Hyoscine N-butylbromide.
Spare sterile gauze squares.
Hibitane/spirit.

EMERGENCY DRUGS.

CHAPTER 7

CYSTOGRAPHY
(*David Edwards, F.R.C.P., F.F.R.*)

THE sub-headings of this chapter are Simple Cystography, The Micturating Cystogram and Stress Cystography. Up to the stage at which contrast medium is introduced into the bladder the points under discussion are similar and will be considered first; thereafter the accounts are given separately.

Contra-indications

1. Acute infections of the bladder or urethra.

2. Known sensitivity to diodone provides an indication for the use of another medium—either barium sulphate (but see 'Contrast Media' below) or a different iodine-containing medium. In the latter case the necessity for the examination should be considered and if it is still thought to be essential the precautions outlined in Chapter 3 should be taken.

3. A urethral lesion which prevents passage of a catheter renders the examination more difficult; ascending injection of contrast or supra-pubic puncture and catheterisation may then be indicated.

Chronic or recurrent urinary infection does not contra-indicate cystography of any type; indeed this investigation is often especially valuable in such conditions.

Preparation.—No special preparation is needed; dehydration is unnecessary. Premedication is not ordinarily required and in the very young may add to the difficulties of the examination. The administration of prophylactic antibiotics before catheterisation can lead to infection with resistant organisms and should therefore be withheld until the examination has been completed—see After-care.

Department preliminaries.—The patient is asked to empty the bladder before catheterisation in the department. When catheterisation is carried out on the ward any residual urine should be measured and a drainage bag should be connected to the catheter; this ensures that time is not subsequently wasted in emptying the bladder in the department.

Procedure

Equipment required.

1. The catheter. This should be of soft rubber or polythene.

2. Apparatus for filling the bladder—either

    (*a*) a glass funnel, length of rubber tubing (18–24 in., 45–60 cm.) and a connection to join it to the catheter; or (*b*) a 50 ml. syringe with a suitable adaptor.

(*c*) a drainage bag *without* a non-return valve can be used as a reservoir of contrast. It is connected to the catheter in the usual way and is then hung inverted from drip stand. It is filled via the small detachable drainage tube using a funnel. Alternatively a drip bottle may be used for administering the contrast.

3. Lignocaine 2% gel with chlorhexidine for urethral anaesthesia in male patients and for lubrication of catheters in the female.

4. In micturating cystography a receptacle is required for the collection of voided urine:—

Female patients: A polythene funnel is gripped between the thighs with a connecting tubing running into a receptacle at the patient's feet.

Male patients: A urine bottle or a plastic jug may be held by the patient or again the polythene funnel method is entirely satisfactory; the funnel is then held in the patient's hand.

Children and infants: Either of the above methods may be satisfactory in older children. In the very young it is preferable either to sit the child in a plastic bucket or on a pot fixed firmly on top of a stool. This means that the lateral projection has to be used but the advantage is that the child's reluctance to void is minimised. To ensure that any reflux which occurs is correctly lateralised the child is rotated into the A.P. position immediately voiding has ceased. In infants it is necessary to examine the child lying in the supine position on absorbent wadding, as noted below.

Contrast medium

Sodium iodide has been widely employed in concentrations of 7–12½% but should be avoided because it may produce a chemical cystitis and is more likely to produce symptoms of iodism.

Barium sulphate made up as a sterile suspension provides a cheap and satisfactory contrast medium. It is used in strengths varying between 20% and 50% depending on the patient's build. It may readily be prepared from powdered barium sulphate; there is also available a much more expensive commercial preparation (Steripaque-Damancy). However barium should be avoided in the presence of a residual urine exceeding 30 ml., when diverticula are present or when vesico-ureteric reflux is suspected. In these cases a barium residue in the kidneys or bladder may act as a nidus for the development of a urinary calculus.

Diodone or Hypaque, which can be purchased in bulk and sterilised by autoclaving (see Appendix) are excellent but more expensive contrast media. Apart from known sensitivity there are no contra-indications to their use. They should be employed whenever barium is contra-indicated. Concentrations from 20% to 40% are suitable, varying according to the patient's size and the method of use, i.e. higher concentrations are required to outline the urethra or ureter satisfactorily than on the occasions when it is desired to demonstrate a filling defect in the bladder. Whatever concentration is used *it is most important to warm it to body temperature before instilling into the patient*.

Catheterisation.—It may be found necessary to arrange for the patient to be catheterised in the ward. If so the ward staff must be instructed to ask the patient to empty the bladder before catheterisation and to record the residual urine, sending this information with the patient. The use of a disposable drainage bag attached to the catheter helps to ensure complete emptying. When infants and young children are catheterised on the ward the excessive use of adhesive tape to hold down the catheter should be avoided as its removal can distress the child and make him unco-operative. The best method is to use a relatively long loop of tape so that the catheter is held away from the patient's skin; the intervening adhesive tape can then be cut with scissors without having to detach it from the skin (see also Fig. 39) The disadvantage of allowing catheterisation in the ward is that responsibility for maintaining asepsis passes from the radiologist and that the catheter is in position for much longer. For this reason it is preferable to catheterise the patient in the X-ray department. Female patients are catheterised by a nurse; the radiologist himself passes the catheter in male patients. In male patients the urethra should be anaesthetised with lignocaine chlorhexidine gel before catheterisation and in all cases the catheter is lubricated with the gel. Full aseptic precautions are essential. Any residual urine is removed by supra-pubic abdominal compression and when a syringe is employed for instillation of contrast, aspiration of the bladder makes emptying more complete. If the urogram has shown marked ureteric dilatation the ureters are emptied by bimanual compression of each loin in turn. The residual urine is measured and the patient is then ready for introduction of contrast.

THE SIMPLE CYSTOGRAM

As intravenous contrast media have improved the indications for this examination have become fewer. It is unreliable as a means of diagnosis in bladder neoplasms because the tumour may be partially or totally obscured by contrast. If radiological methods are to be employed the double-contrast cystogram (Doyle, 1961) gives a more accurate demonstration of the extent of the tumour; even then, more information is normally obtained by cystoscopy. Vesical diverticula may be well displayed by cystography and their presence forms the main indication for the investigation especially in cases where urography gives poor delineation owing to dilution of contrast medium by residual urine. (However, diverticula are usually secondary to obstruction so that unless the nature of the obstructing lesion is known —e.g. prostatic hypertrophy—it is in fact more appropriate to perform a micturating cystogram in order to outline the bladder neck and urethra.)

For this type of examination 150–200 ml. of contrast are normally sufficient and in cases of chronic cystitis or contracted bladder much less may produce uncomfortable fullness. When filling is complete the catheter is attached to the inner aspect of the thigh with adhesive tape and closed with a spigot or artery forceps. The following films are then taken with the overcouch tube (unless there is some special indication for screening).

A.P. with 20–25° caudal angulation of the tube.

Both oblique views with similar angulation.

A Chassard-Lapiné 'outlet view' of the pelvis to demonstrate diverticula arising close to the bladder base. If for any reason this view cannot be obtained an A.P. film with the patient in the lithotomy position will provide a similar projection.

A lateral film is of value in showing bladder displacement or deformity particularly when it is involved by extrinsic masses.

THE MICTURATING CYSTOGRAM

Pathological changes in the bladder may be well demonstrated in the films taken following intravenous urography but the density of contrast medium is not always adequate to demonstrate the bladder neck and posterior urethra. A further drawback of this approach as a method of cysto-urethrography is that it is frequently quite impossible to differentiate between reflux of the bladder contents into the ureters and opacification of the ureters by contrast medium which has been excreted by the kidneys. Distension of the bladder with contrast medium introduced through a urethral catheter or by ascending injection or supra-pubic catheterisation (see p. 61) is the only reliable means of demonstrating the anatomy and physiology of the lower urinary tract. This method, unfortunately, carries the risk of introducing infection, particularly if there is residual urine, and a strict aseptic technique should be followed by administration of sulphonamides for three to four days—see After-care.

Uses of the Method

1. Demonstration of abnormalities in the bladder neck and urethra.
2. The full assessment of cases with vesical diverticula.
3. Demonstration of vesico-ureteral reflux.
4. Demonstration of bladder contractions and the control of micturition.

Reflux of the bladder contents into the upper urinary tract is now recognised as a relatively common cause of renal disease and especially of hydronephrosis and renal failure in children. Other methods of demonstrating vesico-ureteral reflex such as the delayed cystogram are inadequate because although reflux may be shown to have occurred the following important questions remain unanswered.

(i) The presence or absence of obstruction at the bladder neck or urethra.
(ii) The nature of bladder contractions.

Furthermore reflux may only occur during micturition and will be missed if reliance is placed on the delayed cystogram.

The demonstration of bladder contractions is of value in the diagnosis of neurogenic disorders of micturition whilst the peripheral control of micturition may be readily assessed by asking the patient intermittently to start and interrupt the stream. Cine-radiography with an image intensifier is of great value in these cases and may indeed be used as a routine method for cysto-urethrography, particularly for showing vesico-ureteral reflux.

Technique

The patient lies supine on the tilting table with the platform attached so that he can be brought upright. The patient is catheterised as described in the Appendix and any residue in the bladder is emptied. This may require aspiration with a syringe: it is important to produce complete emptying since residual urine may 'layer' above the contrast and will then be the last of the contents of the bladder to be voided. If reflux occurs at this stage it will not be visible. Warmed contrast medium is then instilled until the bladder is filled to capacity. When using the funnel method the funnel is held 2–3 feet (60–75 cm.) above the table. An initial desire to void may be complained of during an early stage of filling, especially if the contrast is introduced rapidly. The patient should be reassured and bladder filling continued until full capacity is reached. The shape of the bladder is helpful in assessing this since the outline when full is at least spherical or may show greater longitudinal than transverse dimension. It is most important to produce adequate filling since underfilling will inevitably cause delay in micturition, or failure to micturate. It is not sufficient to ask the patient if the bladder feels full; objective evidence of discomfort must be present, such as wriggling of the trunk, grimacing and clenching the toes. The bladder capacity varies considerably and may be as much as 1200 ml.; an average amount for an adult is 500 ml. The volume run in should be recorded. Intermittent screening should be commenced early during filling as diverticula or filling defects will be most obvious then. Any reflux seen at this stage should be recorded as it may only occur at this time.

As soon as the bladder is filled to capacity the catheter is removed and the abdomen is screened. Films of the bladder may be obtained, preferably in both oblique positions. If reflux is seen reaching to the kidneys, a film of the whole abdomen may be taken at this stage. The table and patient are then brought into the vertical position, and the receiver for the voided contrast medium is placed in position. A serial film for at least 3 and preferably 4 'spot' exposures should be ready in the film carrier. The patient is rotated into either oblique position under screen control until the region of the bladder neck (frequently represented by a small 'beak' of contrast medium) is well seen. In the left anterior oblique position the male patient should be told to hold the receptacle in the right hand and if the right anterior oblique position is used the receptacle is held in the left hand, so ensuring that an unobstructed view of the urethra is obtained. *The other hand must be kept out of view.*

The patient is then told to empty the bladder and serial exposures are made to demonstrate the urethra and also include the lower ends of both ureters—it is necessary to rotate the patient during the act of micturition so that the bladder neck and urethra are well displayed and both lower ureters included in the exposures. A suggested routine is to start with two exposures in differing degrees of rotation centred over the bladder neck and urethra; then an exposure may be made of the region of the corresponding lower ureter—the right ureter in the right anterior oblique position and vice versa—followed by one of the region of the opposite lower ureter after rotation of the patient. The two latter exposures can usually be centred to show something of the region of the bladder neck. Finally the serial film is removed and a film large enough to obtain an A.P. view of the pelvis, or pelvis and abdomen is taken after completion of micturition. This film shows:

(i) The residual urine in the bladder.
(ii) The extent of any ureteric filling which might have occurred.
(iii) Any diverticula which may have filled during micturition.

When a significant residue is present the patient should be told to go to the lavatory and try to void further. An 'embarrassment residue' will usually disappear. The patient must return for fluoroscopy immediately voiding is completed since occasionally reflux is seen at this stage.

Screening throughout the act of micturition is essential so that transient ureteral reflux is not missed. The nature of the bladder contractions should also be carefully assessed—the normal bladder when emptying resembles a slowly deflating balloon. There is comparatively little time available for obtaining all the films and it is advisable for any radiologist beginning to undertake this examination to familiarise himself with the normal and abnormal appearances so that he knows what he is attempting to achieve. It is desirable to note the presence and force of ureteric peristalsis when reflux occurs since this cannot be shown on still films.

Special problems

Infants and young children.—The very young are best examined in the supine position and assistance or the use of a Bucky band or other form of immobilisation will be required to maintain the child in a satisfactory position. Filling should be accompanied by intermittent fluoroscopy so as to show any moments of reflux. Between moments of screening the perineum is watched, if necessary by an assistant with a torch. In the normal infant micturition occurs involuntarily as soon as full capacity is reached and once initiated will ordinarily continue until completed. The catheter is therefore rapidly withdrawn as soon as the child is seen to void and films are obtained as described above. The degree of bladder filling may also be assessed by abdominal palpation—the full, tense bladder of the child being easily palpable. Very occasionally a short general anaesthetic may be necessary. When sedation is employed it must be adequate since incomplete sedation not uncommonly produces a fractious child resenting any disturbance. With general anaesthesia or when a well-sedated infant stops voiding before complete emptying has occurred it may prove necessary to empty the bladder by manual expression—a gloved left hand being used to compress the bladder through the abdomen leaving the right hand free to manipulate the serial changer.

When infants and young children are examined in the supine position in this way they should lie on absorbent wadding; but care must be taken not to place it too high or the absorbed contrast may partly obscure the bladder and urethra. Children between 18 months and 4–5 years present the greatest problems in management since their fears may make it difficult to reason with them. Premedication with Valium (diazepam) or Librium is often successful in producing a more relaxed attitude. Other factors which are helpful in this respect are a calm atmosphere in the X-ray room, a minimum of waiting and a comfortable, warm table. The presence of the parent with the child is usually a source of re-assurance and the use of sweets such as 'Smarties' can help to create a favourable impression. When screening with an image intensifier it is not usually necessary to turn the lights out and this helps to remove a further source of apprehension. The child should also be re-assured that the X-ray machine is securely supported and will not crush him or her; it has been discovered that this is a common fear among children. Notwithstanding all these points, some children may fail to co-operate and it is then best to employ heavy sedation. Children should therefore be 'booked' for the beginning of a session so that there is time for sedation to be given and to take effect, should it be needed.

Inability to void.—Two groups of patients may be unwilling or unable to void—children and adults. Half-sedated, unco-operative children, particularly girls, may refuse to void. This may give rise to difficulty. The following manoeuvres have all met with partial success: (i) asking the parent, when present, to add his or her persuasion, (ii) warm compresses on the pudenda, (iii) manual expression, (iv) placing a kidney dish under the child so as to simulate a 'pot'.

Adult patients, particularly women, may also be unable to commence micturition, owing to insufficient bladder distension together with embarrassment. If encouragement to relax, together with noisy running of taps does not help the patient she should be sent out and given fluid to drink until she is ready to void. When time is short intravenous frusemide (Lasix) 20 mg. will almost always produce rapid bladder filling and micturition (J.S.M. Beales—personal communication).

Urethral obstruction

The patients with urethral obstruction may be impossible to catheterise.

In these patients a retrograde injection urethrogram will demonstrate the site and nature of the obstructing lesion. It is rarely necessary to fill the bladder in such patients; however, if a cystogram is required the bladder may be filled via the Knutsson apparatus or by a balloon catheter as used for retrograde urethrograms. About 150 ml. of Hypaque 45 or an equivalent contrast medium is injected followed by saline if necessary to produce further distension of the bladder. With this retrograde method, particularly in the presence of urethral obstruction or injury, reflux of contrast medium may occur into the corpora cavernosa and thence into the venous system. *The contrast medium used must therefore be one which is suitable for intravenous injection (Conray, Hypaque, Urografin). Under no circumstances should barium be used and any contrast medium which is not pyrogen-free is quite unsuitable.*

Percutaneous suprapubic catheterisation

An alternative method for performing cystography, either in the presence of urethral obstruction or as an elective procedure, is by percutaneous suprapubic catheterisation (Bryndorf, Christensen, and Sandøe, 1960). A pre-requisite for this approach is that the bladder should be fully distended. Many patients with lower urinary tract obstruction will have distended bladders but otherwise the patient is encouraged to drink freely about an hour before the examination. In infants and young children the bladder is palpable when it is full and puncture is not attempted unless this is the case.

Procedure.—The patient lies supine and the abdomen is palpated to define the bladder, the pubic symphysis and any masses which might be present. In the presence of scars, puncture is made to one side of the scar; otherwise puncture is made in the midline. The skin is thoroughly cleansed with chlorhexidine/spirit (in adults it is usually necessary to shave the pubic hair). A local anaesthetic wheal is raised 1–2 cm. above the pubic symphysis and then an intramuscular needle is attached to the syringe of lignocaine; directing it caudally and posteriorly, at about 45°, it is passed downward until the tissues have been infiltrated. Sometimes the bladder is entered during this infiltration; if so the depth of the needle is noted for the later puncture. The skin at the puncture site is pierced with a no. 11 blade and a 4″ Teflon catheter-needle is introduced along the same path as the local anaesthetic needle. According to the thickness of the abdominal wall the bladder is usually found between 2–3 cm. and 7–8 cm. from the surface. It is advisable to introduce the needle with a series of short, firm thrusts to ensure that the bladder wall is cleanly traversed. A slight snapping sensation or 'give' can often be appreciated as the bladder wall is punctured. The stilet of the needle is removed and if urine emerges the hub of the needle is firmly grasped to maintain its position while the catheter is advanced 4–6 cm. into the bladder. Next the needle is withdrawn and is normally followed by a trickle of urine from the catheter. The catheter is fastened to the skin with adhesive tape and the remainder of the urine is aspirated. Contrast is injected via a flexible connection and filling is watched intermittently on the screen. Voiding takes place spontaneously in infants but in older subjects the routine is similar to that described earlier. The catheter is not removed until voiding is complete, otherwise extravasation may occur. The possibility of minor extravasation or intravasation always exists with this technique and *it is therefore important that the contrast used should be suitable for intravenous injection* and not that prepared from bulk supplies in the hospital.

This is a simple technique and seldom gives rise to difficulty provided the bladder is fully distended.

Neurogenic disorders.—These patients may prove very difficult. Examination in the supine position may be unavoidable. The bladder may be completely insensitive and the degree of bladder distension must then be assessed by abdominal palpation. Bladder contractions may be completely absent; in many cases the patient may completely or partially empty the bladder by contraction of the abdominal muscles but frequently the assistance of manual expression is necessary—this may be done by the patient or by the examiner.

Difficulty in control[1]—stress incontinence.—When a patient complains of incontinence, however infrequent, some modification of technique is needed. Those with enuresis or simple incontinence are made to interrupt voiding as described later in this section. Patients with stress incontinence require, in addition, films to show the position of the bladder base, the size of the cysto-urethral angle and the state of the bladder neck. A 50% suspension of barium is an excellent contrast medium for this later examination provided the contra-indications already discussed are observed. The choice of a suitable urethral catheter is important; a rigid catheter may disturb the relationship between the bladder base and the urethra and it is therefore essential to use a soft rubber catheter. An old rubber catheter which has been boiled many times is very suitable for this purpose. The bladder is filled to capacity[2] with contrast medium and the catheter left in situ; the lumen of the catheter is then opacified with a viscous contrast medium ('Umbradil') and maintained in position by strapping its outer end to the inner aspect of the thigh. Two erect lateral films of the pelvis are then taken with a 'Bucky', the tube being centred 2″ above the greater trochanter; one film is taken with the patient erect and relaxed and the other with the patient erect straining down. It is essential that both the symphysis pubis and the tip of the sacrum are included on these films. A 12″ × 10″ film with its long axis horizontal is generally sufficient but with obese patients and particularly on the 'straining' film, it is safer to use a 15″ × 12″ film. Many patients, because of their fear of incontinence, will not strain sufficiently during this examina-

---

[1] Some patients with 'incontinence have in fact a small capacity bladder and after 100–200 ml. of contrast experience an almost incontrollable urge to void. Usually it will be found that when the bladder is partially emptied the urge passes off and normal control is re-established.

[2] Some workers (e.g. Jeffcoate, 1958) advise against the use of more than 150 ml. of contrast on the grounds that filling to capacity will stimulate detrusor contractions and may abolish the normal urethro-vesical angle. This has not been our experience and we prefer the larger volume to ensure that a satisfactory micturating cystogram is obtained.

tion; it should be explained to them that incontinence may well occur during the investigation, that it is expected and that it should not cause embarrassment. Having obtained these films it is then our practice to carry out a routine micturating cystogram and record the mechanism of control of micturition by asking the patient voluntarily to interrupt the stream during fluoroscopy and filming. This method is used for all types of incontinence and ensures that maximum information is obtained concerning both the state of the bladder and the control of micturition. In particular the presence or absence of 'retrograde emptying' of the urethra above the external sphincter can be confirmed. This 'stripping back' of the urethra is an important normal finding and is essential to 'passive' continence. As already mentioned, a cine film is preferable for demonstrating the control of micturition but adequate diagnostic results can be obtained with a hand serial film changer.

In the past the examination has often been conducted by exposing a third lateral film during the act of micturition rather than performing full micturating cystography. This approach provides a less complete examination of the urethra but it may be preferred when it is not convenient to screen the patient (150–200 ml. of contrast are then adequate). The funnel mentioned earlier should be used for the micturating film. Alternatively the entire examination may be carried out with the patient sitting on a radiolucent—e.g. polythene—bed pan which is placed on a chair. The advantages of this over the standing position are that inhibition of straining and micturition are less likely and that it is easier to maintain the patient's position unaltered for the three exposures.

Complications and After Care.—Infection introduced by catheterisation is the chief hazard. It must be stressed that a scrupulous aseptic technique must be maintained and sulphonamides should be given for four days following the examination, e.g.

1. Sulphamethoxypyridazine (Midicel) 1–2 G. initially, then 0·5 G. daily.
   For children Midicel suspension 1–4 teaspoons daily according to age
or 2. Acetyl sulphafurazole (Gantrisin) 1 G. bd.

References and further reading
Double-contrast cystography
DOYLE, F. H. (1961). *Br. J. Radiol.*, **34**, 205.

Micturating cystography
BRYNDORF, J., CHRISTENSEN, E. R. and SANDØE, E. (1960). *Acta Radiol.*, **53**, 204.
EDWARDS, D. (1960), Cine-radiography of the lower urinary tract, p. 88. *Modern Trends in Diagnostic Radiology* (Butterworths, London).
HODSON, C. J., and EDWARDS, D. (1960). *Clin. Radiol.*, **11**, 219.
KJELLBERG, S. R., ERICSSON, N. O., and RUDHE, U. (1957). *The Lower Urinary Tract in Childhood.* (Year Book Publishers, Inc.)
MCGOVERN, J. H., MARSHALL, V. F., and PAQUIN, A. J. (1960). *J. Urol.*, **79**, 932.

Stress cystography
JEFFCOATE, T. N. A. (1958). *J. Fac. Radiols.*, **9**, 127.
ROBERTS, H. (1952). *Br. J. Radiol.*, **25**, 253.

APPENDIX

Trolley setting

Sterile—upper shelf

2–3 sterile towels. Swabs.

2 small bowls or gallipots, one with soap solution (B.P.) or Betadine for cleaning urethral meatus and surroundings.

Kidney dish to take vesical residue.

Half-litre measure jug.

Sponge holder or artery forceps to clamp tubing.

Funnel,[1] 1 metre rubber tubing and adaptor to fit catheter *or* 50 ml. syringe, 2-way tap and Bardex adaptor *or* funnel and disposable drainage bag to be employed as a store of contrast medium as described earlier.

Catheter spigot.

Surgical gloves. Paper towel for drying hands.

Lower shelf

4 × 100 ml. bottles of 50% Hypaque or Diodone.

500 ml. bottle of sterile water.

Bottles of: Betadine
            Chlorhexidine/spirit

Tube lignocaine 1%/Chlorhexidine gel (sterilised by immersion in chlorhexidine spirit),

Sterile packs: Assorted catheters (for infants the Guy's feeding tube (4·5 F.G., William Warne) is suitable)
              Assorted adaptors
              Xylocaine nozzle
              Dissecting forceps

A small syringe or a needle may be needed to deflate self-retaining catheters.

Also required

6″ plastic funnel with 4′ connecting tubing leading to a bedpan or bucket—for voiding by female patients and children

Urine bottle—male.

Absorbent wadding for infants and young children.

Immobilisation board if used for infants.

Contrast media

Barium sulphate 50% suspension autoclaved (by the hospital pharmacy) and put up in 100 ml. bottles.

Diodone 70% solution is supplied unsterile in 1 litre bottles by May and Baker Ltd., Dagenham. This is made up in 100 ml. bottles as 50% or 60% solution and then autoclaved.

Hypaque is supplied in powder form by Bayer Products Co., Surbiton on Thames, Surrey. It is made up conveniently as 50% solution in 100 ml. bottles and autoclaved.

Technique of catheterisation

Catheterisation, though necessary, carries the risk of infecting the bladder, particularly when there is a residue. It is therefore important to take precautions to maintain asepsis, including the wearing of mask and gloves. The skin of the glans or vulva is cleansed not to achieve sterility, which is impossible, but to remove the grosser contamination. Suitable fluids are Savlon (cetrimide 1% with Chlorhexidine 0·5%), Domiphen/Chlorhexidine (see chapter 4) or Betadine. The anterior urethra, often a source of organisms, is best sterilised with 0·1% chlorhexidine and 1% lignocaine gel; this also makes the procedure more comfortable.

Female patients. A good light is essential. The legs are widely parted and a sterile towel is laid on the table between them. It helps if an assistant is available to pull the labia apart. The labia majora are cleaned with one of the fluids mentioned above, then they are parted with the thumb and forefinger and wiping from before backward further swabs are used, once only, to cleanse the folds between the labia, then the labia minora and finally the urethral orifice. Chlorhexidine/lignocaine gel is squeezed into the anterior urethra and a swab soaked in the cleaning fluid is placed in the pudendal cleft while the catheter is made ready. It is taken in

---

[1] Some plastic funnels have ridges on the outside. These prevent an adequate fit of the tubing and should not be used.

forceps and laid in a kidney dish. This is placed on the sterile towel, the labia are again parted and the catheter, held in forceps, is advanced into the urethra without touching the skin. The catheter should not be passed far into the bladder as this may lead to incomplete drainage and so to layers of unopacified urine above the contrast.

Male patients. The patient should preferably begin by carefully washing the penis with soap and water. It is then swabbed with cleansing fluid, holding it in a sterile swab and is surrounded with sterile towels. The foreskin is retracted and thorough swabbing is continued around the glans and in the urethral meatus. A tube of Chlorhexidine/lignocaine gel is squeezed into the urethra using a xylocaine nozzle; a penile clamp is employed to keep the gel in. A catheter is placed in a kidney dish and after a suitable interval to allow the gel to take effect, is passed into the bladder using forceps.

CHAPTER 8

INJECTION URETHROGRAPHY IN THE MALE[1]
(*David Edwards, F.R.C.P., F.F.R.*)

THE urethra may be examined by the ascending or descending route. The urethra is well displayed during micturating cystography (Chapter 7) and for certain lesions, e.g. urethral valves, this is the only suitable method of demonstration. For investigation of the female urethra it is also the most convenient approach. But for many urethral lesions injection urethrography provides better definition and greater control.

Indications

1. Demonstration of congenital urethral anomalies associated with hypospadias or epispadias, e.g. diverticula.
2. Demonstration of urethral injuries.
3. Investigation of urethral strictures—inflammatory, traumatic or carcinomatous.
4. Demonstration of false passages.
5. Investigation of peri-urethral abscesses and fistulae, prostatic abscesses and cavities.
6. Investigation of prostatic enlargement—inflammatory, benign hypertrophy and carcinoma.

Contra-indications.—The examination should not be carried out in the presence of acute urethritis and balanitis. If there is a history of sensitivity to contrast medium the necessity for the examination should be reconsidered. If it proves absolutely necessary preliminary treatment with antihistamines and hydrocortisone should be carried out (see Chapter 3).

Preliminaries.—No preliminary preparation is required but the nature of the examination should be carefully explained to the patient and immediately prior to the examination the patient is asked to empty his bladder.

Contrast medium.—For the majority of cases a viscous contrast medium is desirable so as fully to distend the urethra and thus demonstrate minor encroachments on its lumen. The most satisfactory medium of this type is Umbradil viscous U., a jelly-like preparation supplied in 40 ml. tubes.

However, in cases of suspected urethral rupture a simple water-soluble medium (e.g. 60% Urografin) is more suitable because:—

---

[1] Injection urethrography in the female is employed by some workers, e.g. Gullmo (1962), but is little used in this country and a discussion of the methods is beyond the scope of this book.

(i) A lower pressure of injection is needed.

(ii) The fluid medium runs more easily into minor tears or lacerations or around blood clots and other obstructing debris.

(iii) No residue—even temporary—is left if extravasation into the peri-urethral tissues occurs.

Equipment needed.—Numerous methods have been employed for the introduction of contrast medium into the urethra. Our own preference is for the use of the Knutsson penile clamp and the main description of technique is based on the use of this instrument. It is, however, unduly expensive for those who only perform occasional urethrograms and two simpler devices will be mentioned with a brief indication of their use.

FIG. 6.—Urethral nozzle. This is also of value in sinography—see Chapter XXIV

FULL SIZE

1. *The urethral nozzle and penile clamp.*—The conical rubber urethral nozzle (Fig. 6) attaches directly to a syringe (preferably 30 ml.) of contrast medium. It is introduced into the urethral meatus and pressed firmly in while the thumb and forefinger of the left hand grasp the penis behind the glans. The syringeful of contrast is injected, the nozzle removed and a penile clamp applied, drawing the penis forward to straighten the urethra. Contrast which has exuded is wiped away and films are taken, whenever possible under fluoroscopic control. The main objections to the use of this method are:—

(i) It is not possible to screen the urethra during the injection of contrast medium.

(ii) The injection pressure can not be maintained during the actual filming and, due to the absence of complete urethral dilatation, minor degrees of narrowing of the urethral lumen may not be detected.

2. *The balloon catheter.*—The urethral meatus is the narrowest portion of the anterior urethra. A balloon catheter introduced into the anterior urethra and gently inflated to grip the walls will not pull out if the degree of inflation is correct. Contrast may then be introduced through the catheter using a suitable adaptor to connect it to the syringe. The inflation is painful unless urethral local anaesthetic jelly is introduced into the anterior urethra about 5 minutes previously. Most urethral balloon catheters are too thick-walled for viscous media but are adequate for use with watery media.

The method may be employed either in conjunction with overcouch techniques or preferably with screening as described below. It is particularly suitable for examination of the urethra in suspected rupture following pelvic trauma. A water-soluble contrast medium suitable for intravenous injection *must be employed*. Using a balloon catheter in this way and injecting up to 200 ml. of 45% Hypaque with appropriate screening and films, the bladder can be filled sufficiently to allow subsequent voiding studies. This is mentioned in the preceding chapter.

3. *The Knutsson clamp.*—This instrument is shown in Fig. 7 and its use is described in the ensuing account.

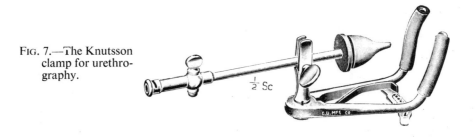

Fig. 7.—The Knutsson clamp for urethrography.

Technique

Having emptied his bladder, the patient lies on the X-ray table. A careful aseptic technique must be observed with thorough cleansing of the glans and external meatus. With the prepuce fully retracted the limbs of the adjustable clamp are made to grip the penis firmly behind the glans. The cannula with its tip protected by a rubber cap is advanced through the external meatus as far as possible into the urethra and is then fixed in position by means of the screw on its supporting arm. The two 20 ml. syringes are filled with contrast medium. Owing to the viscosity of the contrast medium, this is best done from the open end of the syringe with the piston removed—any air bubbles being subsequently expelled. One filled syringe is then connected via the rigid polythene tubing to the end of the cannula.

The patient is rotated into an oblique position with his left side raised, the right leg being flexed at the hip and knee and the right ankle tucked behind the left calf. The fluoroscopic screen or image intensifier is then brought into position by an assistant and the controls covered with a sterile towel so that the operator can manipulate them. The anterior urethra should be straightened by stretching the penis across the right thigh and injection of the contrast medium started. This is accompanied by some discomfort, a burning sensation, which soon disappears. Several 'spot' films of the urethra should be taken in varying degrees of rotation whilst the injection pressure is maintained so that short strictures are not missed. It is essential to obtain both postero-anterior and oblique views of the posterior urethra and bladder neck. The position of the external sphincter can be readily demonstrated by asking the patient to perform the movement of interruption of the stream when contraction of the external sphincter obliterates the lumen of the membranous urethra. This action is normally accompanied by elevation of the urethra—a mobility which may be reduced or abolished by an infiltrating carcinoma or a prostatic abscess.

Filling of the urethra is continued until the bladder base is outlined and this usually requires 30–40 ml. of contrast medium. When changing the syringes, the tap on the cannula should be closed so that contrast medium already within the urethra is not returned. After the spot films have been exposed an oblique view of the whole urethra and bladder base is obtained on 10″ × 8″ film.

When the examination is completed, the clamp is removed and the patient sent out to empty his bladder.

This is a simple technique and rarely gives rise to difficulty. Leakage at the junction of the cannula and the glans may be difficult to control in the presence of congenital anomalies such as epispadias but it is generally due to incorrect placing of the clamp limbs or failure to fix the cannula securely in the urethral lumen. A definite resistance is felt when the head of the contrast medium column reaches the external sphincter but this disappears if the patient is told to relax.

Complications and After Care.—Reflux of contrast medium from the urethra into the venous spaces of the corpus cavernosum not uncommonly occurs, particularly in the presence of urethral obstruction. When marked, it may obscure the urethral lumen and the draining veins may be demonstrated. It does not normally give rise to any ill effects but could produce serious allergic manifestations in the sensitive patient. Dissemination of infection with fever and rigors may follow the examination in patients with peri-urethral or prostatic abscesses and it is our practice to give sulphonamides for four days after the examination (see Chapter 7—After Care). Transient haematuria is common following the examination and the patient should be warned to expect this.

References and further reading

GULLMO, Å. (1962). *Acta Radiol.*, **57**, 71.
MORALES, O., and ROMANUS, R. (1952). *Acta Radiol. Suppl.*, 95.

APPENDIX

Trolley setting

Sterile—upper shelf

2–3 sterile towels.
Sterile swabs.
Bowl of soap solution (B.P.) or Betadine for penile cleansing.
Paper towel for drying hands. Surgical gloves.
Tube of 1 % lignocaine/chlorhexidine gel in a bowl of dilute Dettol or chlorhexidine/ spirit. This is occasionally needed, with nozzle, for anaesthesia of the meatus before urethrography with the Knutsson clamp; it should always be employed before the balloon catheter method.
Instrument for introduction of contrast:—

1. Urethral nozzle (G.U.), 30 ml. syringe and a penile clamp *or*
2. Balloon catheter (5 ml. balloon; 14 F.G.), 50 ml. syringe, Bardex adaptor, polythene connection and 2-way tap.

3. Knuttson's penile clamp (G.U.: Thackray of Leeds) with rigid wide-bore polyvinyl connection. 2×20 ml. double ring grip syringes.

Lower shelf

Contrast medium—Umbradil viscous U.[1] *or* water-soluble media, e.g. Hypaque 45%: whatever medium is chosen *must be suitable for intravenous injection.*
Chlorhexidine/spirit and Betadine.

[1] Supplied by Astrapharm Ltd., Kingston-upon-Thames.

CHAPTER 9

RENAL CYST PUNCTURE

(F. W. Wright, B.M., B.Ch., M.R.C.P., F.F.R.)

SPACE occupying lesions in the kidneys may be cysts or tumours. Renal angiography may show a 'pathological circulation' in 90% or more of renal tumours, but not in all. Direct renal puncture is a reliable method of distinguishing between the two.

Contra-indications.—*Renal tumour.*—While little harm is thought to result from puncture of a tumour, it is clearly undesirable to needle a tumour deliberately and renal puncture is not undertaken unless there is a reasonable likelihood that a mass is cystic. Equally, however, the method is often of greatest value in showing that an apparently avascular mass is in fact a neoplasm.

Hydatid cyst.—Puncture of an hydatid cyst carries a considerable risk of promoting local dissemination: a suspected hydatid cyst should never be punctured.

Special Points in Booking.—It is prudent to admit the patient to hospital for the night following the examination. Ensure that all previous films are available at the time of the examination.

Ward or Out-Patient Preparation.—Bowel clearance and dehydration are instituted as for urography (p. 5).

Department Preliminaries.—The procedure is explained to the patient so as to ensure his co-operation. If the patient is unduly anxious he may be given Valium 5–10 mg. intravenously; ordinarily sedation is unnecessary.

Technique.—The patient lies supine on the fluoroscopy table and a large dose of contrast is injected intravenously. He then turns into the prone position and is settled comfortably with pads or pillows. The operator scrubs up, dons gloves and checks his instruments. After about 5 minutes, when there will be a good urogram, the skin is cleaned and the patient's back and the screening controls are towelled. The kidney is located by fluoroscopy, the beam size is reduced to a minimum, and the site of the lesion is marked by placing the end of the needle on the skin directly above it. Local anaesthetic, e.g. 0·5% Xylocaine with 1/10,000 adrenaline, is infiltrated at this site into the skin and subcutaneous tissues through the muscles of the back to the kidney.

A long needle, e.g. carotid or lumbar puncture needle,[1] is then inserted

---

[1] **Editor's note:** There are considerable advantages in the use of a catheter-over-needle for this procedure, (e.g. Longdwel 4″ or 6″, 18 gauge—catalogue nos. 6719 or 6721). The puncture is made in the usual way and when the stilet is removed from the inner needle, fluid will ooze out slowly; aspiration may be needed to make quite certain that the cyst has been entered. Then the operator holds the needle firmly with one hand while the other advances the catheter down into the cyst; after this the needle is withdrawn. This provides a stable puncture allowing rotation of the patient with little risk of dislodging the catheter. It also gives the best chance of complete aspiration.

into the skin and its position checked by fluoroscopy and parallax to confirm the direction using the smallest possible field. The needle is then advanced vertically downwards towards the kidney. When the renal capsule is reached the needle moves with respiration. Fluoroscopy is again used to check the position and the needle is inserted into the kidney, a slight 'give' being felt as it penetrates the capsule.

If a cyst is entered fluid will exude from the needle; a polythene tube should then be attached and the fluid collected; it is usually straw-coloured but a little blood staining may occur initially. After collection of about 20 to 30 ml., 10 ml. contrast (e.g. Urografin 60%) is injected and the syringe is emptied and filled several times via the polythene tube in order to mix the contrast with the remaining fluid in the cyst. Check by fluoroscopy will show whether the cyst is properly filled or not. Larger amounts of contrast are needed for bigger cysts.

P.A. and oblique films on the Bucky and lateral films with stationary grid and horizontal beam are taken; all the films must be taken prone as the needle remains in the patient's back. When the nature of the lesion is confirmed the cyst contents are aspirated, with a check by fluoroscopy, and the needle is then removed. If 10 ml. of lipiodol are introduced into the cyst before the needle is withdrawn, this provides a permanent marker —the approximate size of the cyst can subsequently be gauged from an erect film.

One must ensure that the cyst demonstrated corresponds exactly with the space occupying lesion, and that it does not contain a filling defect. If it does not represent the whole of the lesion, a second puncture is advisable.

The cyst fluid is examined by a pathologist for malignant cells.[1]

If, instead of being a cyst, the space occupying lesion is a carcinoma, either no fluid exudes from the needle or blood, or bloodstained fluid is obtained. Injection of 1–2 ml. of contrast into the lesion shows a coarse irregularly arranged stroma, which distinguishes it from normal renal tissue. Radiographs are taken to show this.

With large lesions a marker on the skin and a Bucky film may give sufficient localisation for the puncture.

Difficulties and complications.—*Displacement of the needle.*—This may occur when the patient moves, particularly during rotation into the oblique positions: it can also occur as the cyst collapses during aspiration. The likelihood of displacement is diminished if the needle is introduced well into the cyst and if puncture is as nearly central as possible.

Escape of contrast.—This is usually due to injection of larger volumes than have been aspirated. It is not of significance except that it may cause slight discomfort.

Spread of tumour cells.—(i) into the puncture site. Lindblom did not have any cases (personal communication).

---

[1] *Editor's note*: Examination of the fluid for fat has also been recommended (Lang, 1966). Fluid from a benign cyst is fat free and the presence of fat is strong evidence in favour of neoplasm.

(ii) Into the circulation. A theoretical risk, but the danger is probably less than with firm palpation.

Haemorrhage.—As the procedure is carried out under fluoroscopic control, the danger should be minimal.

Pneumothorax.—With a lesion of the upper pole it is better to puncture the kidney substance with the patient holding his breath in expiration, this will avoid traversing the lowest limit of the pleura.

After Care.—It is prudent to keep the patient in hospital for the night after examination.

References

HARE, W. S. (1961). *J. Coll. Radiol. Aust.*, **5**, 68.
LANG, E. K. (1966). *Radiology*, **87**, 883.
LINDBLOM, K. (1946). *Acta Radiol.*, **27**, 66.

APPENDIX

Trolley settings

Sterile—upper shelf

Gloves and powder. Paper towel.
4 towels.
1 syringe 10 ml.
1 syringe 20 ml. (for contrast).
3 needles for local anaesthetic.
1 carotid needle and/or lumbar puncture needle or Longdwel 6″, 18 G catheter-over-needle (B.D. No. 6721) with polythene tube and adaptors.
1 gallipot for antiseptic.
1 gallipot for cyst fluid.
1 forceps to hold swabs.
1 small forceps for handling needles, etc.

Lower shelf

Local anaesthetic.
   Contrast medium.
   Bottle for cyst fluid to be sent to
      Path. Lab.
   Masks.
   Antiseptic.

CHAPTER 10

PERCUTANEOUS PYELOGRAPHY AND PERCUTANEOUS NEPHROSTOMY

PERCUTANEOUS pyelography involves the puncture of an obstructed pelvi-calyceal system and injection of contrast to demonstrate the level of the obstruction. The indications for the examination are few because in most cases retrograde pyelography is equally informative and retrograde catheterisation may provide adequate drainage of the obstructed system. Moreover the use of high dose urography with tomography has proved an even simpler way to confirm the presence of obstruction in the anuric patient. However percutaneous pyelography is a necessary preliminary to percutaneous nephrostomy—a new method of drainage of an obstructed kidney (Ogg, Saxton and Cameron, 1969)—and the two techniques will be described together.

Among the indications for the use of percutaneous nephrostomy are the following:—

1. Short-term relief of obstruction in the anuric patient in whom retrograde catheterisation is impossible. This may be because of technical difficulty, because the condition of the patient precludes anaesthesia or because the ureter is not available for catheterisation, e.g. after implantation into an ileal loop.
2. Short-term relief of obstruction in a patient who is awaiting pyelo-plasty and in whom there is a risk that a pyonephrosis may develop after retrograde catheterisation.
3. Short-term relief of obstruction in a patient with unilateral 'non-function'. It may then be possible to determine the extent to which renal function would improve with permanent relief of the obstruc-tion, e.g. by pyeloplasty.

Summary of the procedures

Percutaneous pyelography

1. A high dose of intravenous contrast (e.g. 100 ml. Conray 420) is given and when the kidney is adequately visualised the patient lies prone.
2. Under local anaesthesia a fine needle is introduced, with fluoroscopic control, into the pelvi-calyceal system and contrast is injected.

Percutaneous nephrostomy

1. After visualisation of the upper urinary tract by percutaneous

pyelography, a wide-bore needle is introduced into the distended pelvis or a calyx.

2. A ureteric catheter is passed down the needle and advanced into the pelvis.
3. The needle is withdrawn, the catheter is fastened to the skin and allowed to drain into a collecting bag.

Equipment required. The items of particular importance are:—

1. Fine exploring needle—similar to that used before renal biopsy. A longer needle (up to 15 cm.) may be needed in heavily built or oedematous patients.
2. Vim-Silverman needle or needle of similar length and bore. It must be wide enough to allow the passage through it of:—
3. No. 5 (French) whistle tipped ureteric catheter. This should preferably be of a 'woven' construction to lessen the chance of obstruction if it becomes kinked. However, plain polythene catheters can be used.
4. Tuohy type of adaptor with serrated hub. This is used to link the catheter to the drainage bag.
5. Alternatively a catheter-over-needle may be employed, preferably with a 'pigtail' type of catheter having a 'built-in' curve at the end (B.D.).

Ward preliminaries.—Sedation is often unnecessary as uraemic patients are frequently drowsy. Diazepam (Valium) 5 mg. i.–v. is usually adequate if sedation is needed.

Percutaneous pyelography and nephrostomy

Visualisation of the kidney

The diagnosis of obstruction is usually made by high dose urography with tomography. This will frequently show early calyceal opacification but sometimes only a 'negative pyelogram'. It may be necessary to wait for an hour or two before the kidney is visible on the screen and sometimes only the nephrogram can be seen. If there is obstruction by an opaque stone, this may provide a sufficient guide as to the correct site of puncture. Should fluoroscopic delineation be inadequate, two further measures are possible:—

1. Markers can be placed over the approximate position of the kidney and a prone overcouch film taken. This will usually give sufficient indication of the best position for puncture. Before skin preparation the skin is marked indelibly at the chosen site.
2. Retroperitoneal nitrous oxide insufflation by the translumbar route (see page 166) will outline the kidney. The exploring·needle is then directed at the estimated position of the renal pelvis. A pad under the abdomen is needed to prevent the kidney 'falling' anteriorly when surrounded by gas.

Puncture of the pelvi-calyceal system

After the preliminaries outlined above, the patient is settled as comfortably as possible in the prone position. Abundant 'padding' is advisable as many patients become restless when lying prone for any length of time. The skin over the appropriate side is cleaned with hibitane/spirit and the back and hips are covered widely with sterile towels. A marker—e.g. a scalpel blade—is placed over the estimated site of puncture and its position is adjusted under screen control; it is advisable to relate the skin puncture to the position of the kidney at a given phase of respiration, e.g. in expiration. Then the skin and deeper tissues are infiltrated with Lignocaine 1% and while this is taking effect a 20 ml. syringe is filled with contrast—e.g. Conray 280—and attached to a flexible connection. The skin is 'nicked' with a scalpel blade and the exploring needle is advanced vertically downward with further injection of Lignocaine and with repeated fluoroscopic verification of its direction. As it nears the region of the kidney, a minor degree of respiratory swing becomes apparent, and once the needle is in the kidney the swing is more marked and can be seen on the screen.

Each time it is thought that the needle is about to enter the pelvis, the patient is instructed to stop breathing and the needle position is checked. The stilet is re-inserted and the patient again stops breathing as the needle is advanced into the kidney. Acute obstruction results in a high pelvic pressure and when the stilet is removed the urine wells out. With chronic obstruction the urine may not be under pressure and it may be necessary to aspirate by way of a flexible connection in order to confirm that the pelvis has been entered. In fact, several attempts may be needed before successful puncture is achieved.

Once the needle is in place, the flexible connection is attached and contrast is injected. 10–15 ml. are sufficient as a rule unless the pelvis is very large; it may be helpful to aspirate and inject several times to produce full mixing. When the pelvi-calyceal system is well opacified a film may be taken; then the wide-bore needle and the catheter are made ready. It should be possible to pass the catheter smoothly through the needle; while doing so it is useful to note the marking on the catheter which corresponds to the moment when it emerges from the needle tip. Both at this stage and when the needle has been introduced into the patient the catheter should not be withdrawn through the needle; to do so risks cutting the catheter on the sharp bevel.

With these preparations complete, the exploring needle can be withdrawn, carefully gauging how far it has been inserted. This distance is measured against the wide-bore needle so as to obtain an estimate of the depth to which the latter must be introduced. Using screen control, as before, and with the stilet inserted, it is then passed to a point just short of the kidney. The patient holds his breath, the needle is firmly advanced into the pelvicalyceal system and the stilet is removed. When it is clear that the needle is correctly placed, usually by a gush of urine, the ureteric catheter is passed gently through the needle and advanced 5–10 cm. into the pelvis. There is normally only a slight resistance as the catheter meets the opposite wall of the pelvis and before it is deflected into the pelvis. Once its position is established urine begins to drip from its outer end and the needle can be withdrawn.

The catheter is now secured by a skin stitch and by firm strapping with adhesive tape, taking care not to kink it. The adaptor is screwed tightly on to the free end and this is joined to the tubing of the plastic collecting bag. Adhesive strapping is liberally applied to prevent accidental disconnection or withdrawal of the catheter. The patient may then return to the ward.

Alternative method.—An alternative method may be mentioned as it seems likely to replace the technique outlined above. As yet insufficient experience has been gained to be sure that it is wholly satisfactory but its advantage is that a wider catheter is introduced, allowing improved drainage. A catheter-over-needle is used instead of the Vim-Silverman needle. A suitable system is modelled on the 'pigtail' catheter-needle evolved for suprapubic cystostomy (B.D.) since this has an inbuilt curve over the terminal portion, together with a number of side-holes for drainage. At the time of writing discussions are being held to try and evolve a range of such needle-catheters.

The catheter-needle is introduced into the renal pelvis as described above. The stilet is withdrawn to confirm that the tip lies in the pelvis; then the needle element is held while the catheter is advanced down into the pelvis. When the needle is withdrawn urine should pour from the catheter. If there is any difficulty a flexible connection is attached and a little contrast is injected to determine the position of the catheter and the reason why it is not draining. Once satisfactory drainage is established the catheter is strapped to the skin and connected to a drainage bag.

Although a catheter with an inherent curve has obvious advantages in that it will pass more readily into the renal pelvis, it is also possible to use a simple catheter-needle in the same way. After puncturing the pelvis the needle is withdrawn and if urine emerges from the catheter a Seldinger 'J'-guide (see p. 235) is passed through the catheter and, with gentle probing, advanced into the pelvis. The catheter is then advanced over it and after checking the position, the guide is withdrawn; drainage is less reliable by this method.

Special problems

Difficulty in puncture.—The causes are:—(a) poor visualisation, making direction of the needle difficult. (b) unusual depth of the kidney— requiring a longer needle for exploration and for final puncture; 15 cm. may be needed (c) thickening of the renal pelvic contents—e.g. in pyonephrosis. (d) a plug of tissue may block the needle if it is advanced without a stilet.

(e) puncture may be impossible if there is no obstruction and hence no pelvicalyceal distension.[1]

Bleeding.—Puncture of a vessel may occur during exploratory needling but the needle is very fine and is unlikely to cause significant bleeding. The wide-bore needle or catheter-needle can usually be directed more accurately and so bleeding is less likely; we have, however, seen blood staining of the urine draining through the catheter but this did not affect the success of the drainage.

Extravasation of contrast.—During acute obstruction there may be escape of urine from the pelvi-calyceal system, commonly by rupture of a calyx into the renal sinus. After renal puncture and introduction of contrast, a very confusing picture can result, but provided the calyces or pelvis are still visible, this need not hinder the technique. It should be made clear that the extravasation is not a complication of the procedure but the result of acute obstruction.

Difficulty with introduction of the catheter.—This is a possible hazard which has not yet been encountered. The following manoeuvres are suggested:—(i) check that urine is still emerging from the needle; if not, advance or withdraw the needle until the flow is re-established. (ii) if the catheter still will not pass, replace the stilet and screen to determine whether a particular phase of respiration or gentle angulation of the needle might dispose the needle more favourably.

Difficulty in drainage.—This may result from kinking or poor positioning of a ureteric catheter and it may not be possible to improve it except by gentle withdrawal. Syringing of the catheter should be tried since this often restores drainage. The narrow lumen of the ureter catheter is one reason for favouring catheter-over-needle systems.

Excessive diuresis.—The sudden relief of obstruction in a uraemic patient allows a vigorous osmotic diuresis and it is important to make good the fluid loss in the period following the nephrostomy.

Aftercare

Apart from maintaining fluid balance no special care is required. Removal of the catheter, when necessary, does not give rise to more than slight discomfort.

Reference

OGG, C. S., SAXTON, H. M. and CAMERON, J. S. (1969). *Br. med. J.*, ii, 657.

APPENDIX

Trolley setting

Top shelf

3 towels; gauze swabs
2 gallipots

---

[1] This account is based on the presumption that a firm diagnosis of obstruction is made before puncture is attempted, but it is theoretically possible to use this technique as a means of determining the presence or absence of obstruction, as in percutaneous cholangiography.

2 × 20 ml. syringes
I × 10 ml. syringe
Pentothal mixer
No. 11 disposable scalpel blade
Novex 3-way adaptor with flexible connection
Gloves. Paper hand towel
Needles: Nos. 1 and 17 hypodermic—for local anaesthetic
       Fine exploring needle
       Vim-Silverman needle *or*
       Bonanno type 'pigtail' catheter (B.D.) *or*
       Catheter over needle (B.D. catalogue nos. 6719, 6721, 6734 or 6735) with
          1 or 2 side-holes and a 40 cm. 'j'-guide.
Linen suture and needle. Needle holder. Scissors.
3/4 adaptor with serrated hub (A.H. 63376)

Lower shelf

Masks: Disposable catheter drainage bag.
Ampoules: Normal saline—for flushing needles and connections
Lignocaine 1% 2 × 5 ml.
Contrast—Conray 280 or Hypaque 45
Sterile pack: No. 5 whistle tip ureteric catheter, preferably woven nylon.

N.B. Contrast for intravenous injection will be needed if it has not already been administered.

CHAPTER 11

HYSTEROSALPINGOGRAPHY
(*Ellis Barnett, D.M.R.D., F.F.R.*)

EXAMINATION of the uterus and its adnexae by hysterosalpingography is not a new procedure, although many variations of technique have been described since Nemenov first outlined the uterus with Lugol's Iodine in 1909. Hysterosalpingography has gradually become accepted as a valuable and often essential part of the investigation of many gynaecological conditions.

Contra-indications

1. Hysterosalpingography is contra-indicated in *the week preceding* and *that following menstruation* when the endometrium is very thick or has been denuded. The reason for this is twofold. Firstly, venous or lymphatic intravasation is very liable to occur near the time of menstruation, and, although intravasation with water-soluble contrast media is not dangerous, this occurrence is nevertheless undesirable and must be regarded as a complication of the procedure. Secondly, if the examination is delayed until some days after menstruation, one is less likely inadvertently to perform a hysterosalpingogram in the presence of an early pregnancy.
2. *Acute vaginitis* or *cervicitis* is a contra-indication because of the danger of ascending infection.
3. The presence of *uterine malignancy* or *suspected malignancy* is widely regarded as a contra-indication to hysterography, as it is claimed that dissemination of malignant cells may occur. Although this theoretical hazard cannot be wholly disregarded, there is no irrefutable evidence to support this postulation. We consider it improbable that hysterography, carefully performed, would be likely to cause more rapid dissemination of a uterine malignancy than would otherwise occur and it must be remembered that since negative curettage may occur in the presence of uterine malignancy, any procedure which might facilitate diagnosis should be considered.
4. *Early pregnancy* is an obvious and absolute contra-indication.
5. *Suspected ectopic pregnancy* with acute symptoms is a contra-indication because of the danger of tubal rupture.

Special points when procedure is arranged.—It follows from the foregoing remarks that a hysterosalpingogram should be arranged about the middle of the menstrual cycle and certainly at least seven days after menstruation. In fact, when the primary aim is the assessment of tubal patency, the best time to carry out the examination is on the day of ovulation. *It has been shown that a higher percentage of tubal filling is obtained on the day of ovulation than at other times.* This day is suggested by noting a rise in rectal temperature. This method is perhaps inconvenient as a routine procedure, but should certainly be adhered to when repeating the examination after one inconclusive result.

General anaesthesia is not necessary in the majority of cases. Usually a simple explanation of the procedure is all that is required. A highly nervous patient may be given 50 mg. pethidine by mouth half an hour before the examination, with good effect. It has been noted that a general anaesthetic often fails to relieve utero-tubal spasm and even the thought of an anaesthetic may induce spasm which is not relieved by anaesthesia.

Inquiry should be made into the menstrual history, particularly with regard to regularity and duration. An abnormal period either in time or duration should be regarded with suspicion. For example, a period of unusually short duration may suggest a false period in association with an early pregnancy. A clinical history should also be obtained with particular reference to intra-abdominal inflammation and tuberculosis.

Ward or out-patient preparation.—Usually it is unnecessary to admit a patient to hospital for a hysterosalpingogram but when a repeat examination is desired, or if the patient is particularly nervous, she should, ideally, be admitted overnight. Hospital admission should also be arranged when there is a recent history of pelvic inflammation, to enable suitable antibiotic cover to be established.

A laxative, e.g. Dulcolax tabs. 2, is taken by out-patients on the evening preceding the examination. In-patients should be given a simple soap and water enema instead, on the morning of the examination, but should not be shaved and prepared as for an abdominal operation, since every effort must be made to avoid inducing tension in these cases, especially when the investigation is for infertility. As mentioned above, nervous patients should be given 50 mg. pethidine by mouth half an hour before the examination, and should be allowed to relax quietly thereafter. This applies both to in-patients and out-patients.

Departmental preparation.—A tray or trolley is prepared holding all the items necessary for hysterosalpingography—see Appendix.

The Contrast Medium.—Two main groups of contrast media are available at present, namely oily and water-soluble. The ideal contrast medium should have the following properties:—

1. It should be non-irritant.
2. The viscosity should permit ready flow and ease of handling, but the flow should not be rapid at the expense of contour detail.
3. Radio-opacity should be adequate, although extreme density is undesirable.
4. The medium should mix readily with other fluids.

5. It should be readily absorbed within a reasonable time without leaving any residue which might initiate a foreign body reaction.

Oily Media.—In the earlier days of development of hysterosalpingography oily media, e.g. Lipiodol, were used exclusively. The great advantage of this medium is that it adheres well to the mucosa, permitting a number of films to be taken in different positions without loss of contour, and peritoneal spill does not usually induce discomfort. One disadvantage is the slow permeation through the tubes, necessitating a follow-up examination at twenty-four hours to assess tubal patency. This type of medium has been widely condemned because of reports of oil embolism and of foreign body reactions produced by unabsorbed oil in the peritoneal cavity. However, the fact that many workers have used Lipiodol in large numbers of cases without untoward reaction cannot be disregarded when considering the choice of contrast medium and Lipiodol (usually in the form of fluid Neo-hydriol) still has its advocates for general or for limited use. (For a fuller discussion see the papers by Brown *et al.* (1949), Freeth (1952), and Barnett (1955, 1956).)

Water Soluble Media.—There are a number of such media in general use in this country at the present time, namely:—

Endografin (Aqueous solution of the methyl glucamine salt of N.N'-adipic-di-3-amino-2.4.6.-tri-iodobenzoic acid in 50% concentration).

Urografin 76% (A mixture of the sodium and methylglucamine salts of N.N'-diacetyl-3-5.-diamino-2.4.6.-tri-iodobenzoic acid).

Salpix (54% solution of sodium acetrizoate with polyvinyl-pyrrolidone).

Diaginol Viscous (40% solution of sodium acetrizoate with dextran).

Endografin and Urografin achieve their viscosity from the size of the molecule, whereas the viscosity of Salpix and Diaginol Viscous is attained by adding a suitable vehicle, either polyvinyl-pyrrolidine or dextran.

Water soluble media are more easily handled than oily media; they flow more readily and peritoneal spill is usually seen at the time of the initial examination. However, adherence to the mucosa is not as satisfactory as with Neo-hydriol, and contour is soon lost, unless the injection flow is sustained while films are taken. Also water soluble media tend to be rather more irritant to the peritoneum than oily media, although they are absorbed within several hours leaving no residue, and, should venous intravasation occur, there is no danger of embolism.

Although the ideal contrast medium has yet to be discovered, for practical purposes any of the water soluble media mentioned above are suitable for hysterosalpingography. There is little to choose between these media with regard to radio-opacity and viscosity as judged on the X-ray screen. The main factor which will influence the choice of medium is the incidence of pain induced, but reports in the literature are conflicting in this respect (Czyzewski, 1956; Davies *et al.*, 1957; McCann *et al.*, 1957;

Reiss *et al.*, 1958; Sheach, 1959; Henry *et al.*, 1960). All these media can be recommended, and the final choice will depend upon personal preference and experience. Our own preference is for Urografin 76%.

In many centres only water soluble media are used for hysterosalpingography and hysterography and as a general rule this is correct. However, we consider that oily media should not be discarded entirely. We still prefer Neo-hydriol for hysterography when a congenital uterine anomaly is suspected or when uterine distortion or localised weakness of the anterior uterine wall is thought likely after Caesarean section. In cases of suspected intra-uterine tumour one of the newer water soluble contrast media may be preferred to obviate the possibility of obscuring a small intra-uterine

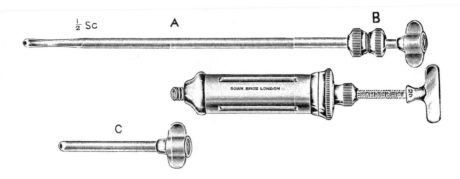

FIG. 8.—The Green-Armytage syringe and cannula for hysterosalpingography. Above is the injection cannula used with a rubber cone (see Fig. 6) for plugging the os. Below are the filling cannula and the syringe with screw action plunger.

filling defect by the extreme radio-opacity of Neo-hydriol. However, such tumorous conditions as adenomyosis can usually be demonstrated very adequately with Neo-hydriol and any disadvantage of radio-opacity may be overcome, if desired, by a double contour technique, the oil being aspirated after filling the uterus, and replaced by CO_2. Apart from the foregoing it is advisable to use a water soluble medium.[1]

Special Equipment

1. *The injection cannula.*—Various types of injection cannula are available. We use the Green-Armytage type (Fig. 8). The rubber cone is placed fairly near the tip of the cannula to facilitate examination of the cervical canal. The flanged screw type of plunger on the syringe is so designed as to eject 1 ml. of the contents of the syringe for every complete turn. This

[1] We have also used Neo-hydriol on occasions for assessment of tubal patency, where both tubal insufflation and a hysterosalpingogram using water soluble medium have given an inconclusive result. Two ml. Neo-hydriol injected into the uterus and retained by a cervical cap may permeate through the tubes and show peritoneal spill in 24 hours.

enables the operator to control exactly the amount of contrast medium injected, even when screening, and also makes the actual injection very simple. The Leech-Wilkinson type of cannula, with a conical metal ridged extremity, is still widely used but this instrument is more traumatic to the cervix; moreover, any traction must be applied very carefully lest the grip on the cervix is lost.

A suction type of cannula originally described by Malmström and his colleagues (Malmström-Westman cannula or VUC) has now become available in this country (Wright, 1961). The cup of the cannula is placed over the cervix and is held in position by suction (Fig. 9). This apparatus is advantageous in so far as it is not traumatic to the cervix but rigid adherence to the recommended pressures is essential. Controlled traction can be applied readily and the degree of penetration of the cannula can be controlled so that the cervical canal can be examined as well as the uterus and tubes. Some difficulty may be encountered with this type of instrument in the presence of a lacerated cervix or a cervix placed very deep in the vagina and directed markedly backwards. Although the basic principle of this instrument is interesting, in practice we have found it much more difficult to use than the Green-Armytage type of cannula.

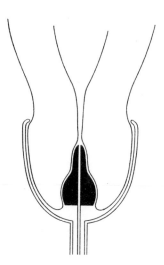

FIG. 9.—Diagram to show the mode of action of the VUC cannula. The cone fits into the cervix while the suction cup fits around it.

2. *The Cold Light Speculum.*—This type of vaginal speculum is particularly useful for hysterosalpingography. The instrument is illuminated diffusely along its whole length. This greatly facilitates insertion of the cannula while also permitting direct visualisation of the cervix and vaginal mucosa.

3. *Image Amplification.*—The use of image amplification in conjunction with a television monitor is also an advance in the technique of hysterosalpingography. Apart from the reduction in radiation dosage, we have been impressed by the reduction in the incidence of utero-tubal spasm since using this apparatus. The patient is encouraged to watch the procedure on the television monitor, and this seems to have a profound effect upon spasm. Although this is merely a personal impression it does tend to support the contention that utero-tubal spasm is functional in origin. The procedure may be recorded on cine-film instead of single cut film.

Double contrast apparatus.—An interesting type of injection apparatus has been developed by Weisman (1952) to produce a double contrast effect. 3-4 ml. of Lipiodol are injected using a special apparatus which allows the operator to change from oil to gas injection without disturbing the cannula, the oil being then forced through the tubes by CO_2.

Departmental Preliminaries

1. The patient empties her bladder immediately before entering the X-ray room. This is of the utmost importance as a full bladder displaces the uterus, interfering with adequate delineation of the uterus and hindering tubal filling and peritoneal spill.
2. Also before the X-ray room is entered the radiologist should explain the procedure in simple terms, emphasising that the investigation is a special type of X-ray examination and not an operation. This is particularly important when dealing with cases of infertility. The patient is warned that she may feel some discomfort during the examination, but should not experience severe pain.
3. A preliminary film of the pelvis is taken anteroposteriorly with the overcouch tube. This control film may give valuable diagnostic information prior to the hysterosalpingogram, e.g. there may be evidence of a calcified tuberculous pyosalpinx, calcified ovary or fibroids, or there may be evidence of bone metastases.
4. The patient is asked to flex her knees, keeping the heels together, and then allow the knees to fall apart. It is not necessary for the true lithotomy position to be adopted. A sterile towel with a suitable cutaway, permitting access to the vagina, is placed over the lower abdomen. A bimanual examination of the uterus and its adnexae may now be carried out by the gynaecologist. The buttocks are then elevated upon a large sandbag, the vulva is swabbed with dilute Dettol and the cold light speculum is inserted gently into the vagina which is then viewed under direct vision and any discharge noted. The vagina is now swabbed with dilute Dettol, the anterior lip of the cervix is gripped by volsellum forceps and the external cervical os touched with concentrated Dettol. A uterine sound is passed to determine the length and direction of the uterine cavity.

The patient is now ready for hysterosalpingography.

Technique

The patient has been positioned as already indicated. The volsellum forceps have been applied to the anterior lip of the cervix, the convexity of the curve of the forceps directed towards the anterior vaginal wall. Too often the forceps are applied with the convexity of the curve directed towards the posterior vaginal wall and this may interfere with insertion of the cannula. The injection cannula containing 20 ml. of water-soluble medium or 10 ml. of Neo-hydriol (according to the indication) is examined and all air bubbles evacuated. (These amounts of contrast medium are, of course, in excess of what is normally required. In actual practice one rarely uses more than 15 ml. water-soluble medium in assessment of tubal patency and rarely more than 4–5 ml. of Neo-hydriol for hysterography. But it is preferable to have more contrast medium in the syringe to allow for any unforeseen loss, e.g. due to a lax or lacerated cervix or perhaps a large

D

capacity uterus holding more than the usual amount of medium.) The injection cannula is now inserted into the external cervical os. Gentle traction is applied to the forceps while pushing gently upwards upon the syringe. This manoeuvre ensures a water-tight junction as far as possible. The sandbag under the buttocks is then removed and the patient allowed to adopt a more comfortable supine position with the legs wide apart and the knees extended. A lead-lined protective box (Fig. 10) is placed between the thighs, the operator inserts the forceps and cannula through the slits provided and the lid of the box is placed in position. The injection of medium is commenced, the operator counting aloud as each ml. is injected (one complete turn of the flanged end of the plunger equals one ml.

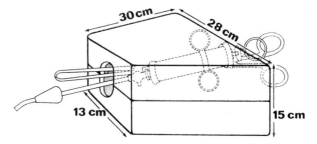

Fig. 10.—Protective box for hysterosalpingography. The lead equivalent should be 0·5 mm.

injected). The flow of contrast medium into the uterus and tubes is observed under screen control using a small field and intermittent screening. (A lead letter R or L is stuck to the screen.) The operator now applies gentle traction to the forceps and cannula together under screen control. This manoeuvre will aid examination of the uterine contour. Using Neo-hydriol no attempt is made to fill the tubes, but when the tubes fail to fill readily with water soluble medium, alternate relaxing and pulling upon the instruments—the so-called 'butterflying manoeuvre'—will encourage tubal filling and also give information concerning uterine mobility.

When tubal filling is observed on the screen the injection should be continued until free peritoneal spill commences. The first film is taken postero-anteriorly at this stage using the undercouch tube. The exposure is made while continuing the injection of contrast medium to avoid any possible loss of contour. In the average case this completes the first stage of the examination. The instruments are then removed. A follow-up over-couch film is taken after about 15–20 minutes to confirm the free peritoneal spill and to show emptying of the tubes. When an oily medium has been employed a film is taken after 24 hours instead to show whether peritoneal spill has occurred.

It may sometimes be evident on screening that a lateral or oblique film is desirable; this should be taken immediately after the first film, the instruments being retained in position and the film taken while further medium is being injected.

SPECIAL PROBLEMS

A. Pelvigraphy.—When there is any clinical suspicion of a pelvic mass pelvigraphy may be performed. The syringe is detached from the cannula after the first stage is complete, but before so doing, the contrast medium in the uterus is aspirated. A second syringe containing 15–20 ml. of water-soluble medium mixed with 20 ml. $\frac{1}{2}$% procaine or lignocaine is connected to the cannula. This mixture is injected steadily under screen control and is allowed to spill freely into the peritoneal cavity. The instruments are then removed and the patient placed on each side in turn to encourage spread of the medium. Films are taken as follows:—

1. An antero-posterior film of the pelvis.
2. One lateral decubitus film using a horizontal beam and another lateral film using a vertical beam, the patient lying on each side in turn. The patient remains on the X-ray table until these films are viewed since occasionally further views are required, e.g. a supero-inferior view of the pelvic inlet or a lateral view of the pelvis using a horizontal beam, the patient sitting in the semi-reclining position.

This procedure is useful as a further stage of hystero-salpingography, but is, of course, dependent upon tubal patency.

B. Cornual obstruction.—Should the tubes fail to fill and the uterine cornua appear rounded, cornual occlusion is suspected. But any variation of cornual shape observed on screening suggests cornual spasm as the cause of non-filling of the tubes. As a rule one examination should not be accepted as conclusive evidence of cornual occlusion. The findings should be confirmed by a repeat examination on the day of ovulation, giving antispasmodics for several days previously (e.g. Buscopan 0·01 G. t.d.s.). Should the tubes fill partially or wholly, but no peritoneal spill be noted either initially or on the routine 20-minute film, a further film about one hour later is of value to confirm the abnormality of the tubes.

C. Technique for examination of the cervix.—In recent years there has been a growing understanding of the importance of incompetence of the uterine cervix as a cause of habitual abortion and it would therefore be appropriate at this juncture to mention the technique of examination of the uterine isthmus and cervical canal, although a detailed description is beyond the scope of this chapter.

Cervical competence depends upon the integrity of the uterine isthmus, which is a transitional area between the cervix and corpus uteri. The isthmus is bounded by the superior isthmic sphincter above, and the inferior isthmic sphincter below, and is about 8–10 mm. in length. It is this short segment which we wish to study particularly in cases of suspected cervical incompetence.

Until recently little attention was paid to the cervical canal during hysterosalpingography, either because the cervix was obscured by the instrument used or was poorly visualised as a result of bad technique.

When using the Green-Armytage type of cannula, the commonest error is to place the rubber cone too far away from the tip of the cannula, thus permitting the tip to project too high into the uterine cavity. In cases of suspected cervical incompetence, the cone should be placed not more than 7–8 mm. from the tip of the cannula. As vaginal reflux will occur more readily under these circumstances, particular care must be exercised to maintain a watertight junction by upward pressure upon the cannula combined with traction upon the tenacula. We have found this method quite satisfactory for routine examination of the isthmus and cervical canal. The VUC cannula, although more difficult to use, is particularly useful for examination of the cervix. The Leech-Wilkinson instrument is liable to distort the cervix and is not recommended. However, it must be accepted that these methods will only demonstrate the more gross examples of cervical incompetence, and moderate or minor degrees may be missed. Therefore for a more accurate assessment of cervical competence, special techniques designed for this purpose are necessary. Probably the most popular basis of these techniques is to utilise a latex balloon attached to a catheter or cannula. The balloon is inserted into the uterus and contrast medium is introduced into it until the uterus is completely filled. Attention is paid to the pressure required to distend the isthmus and cervical canal, or traction may be applied to the cannula while gradually releasing the contrast medium contained in the balloon, until the balloon bulges into the cervical canal through the isthmus. Serial films are taken at intervals throughout the procedure. It is, however, extremely difficult to relate the degree of cervical incompetence with such factors as the amount of traction or the degree of emptying of the latex bag. A preferable modification described by Mann (1959), involves the use of a unit consisting of a latex balloon, joined by a latex neck of varying length (as judged by the length of the uterine cavity) to a hard rubber catheter. The unit is inserted into the uterus by means of a stylet. Injection of contrast medium first produces distension of the distal balloon which lies within the uterine body. As the injection proceeds, the pressure inside the balloon rises until a critical pressure is reached, whereupon the intra-isthmic portion distends, the pressure in the balloon falls, and the isthmus and cervical canal are outlined. According to the critical pressure required, the degree of cervical incompetence can be assessed. This method permits detection of degrees of incompetence which would otherwise be missed by conventional methods.

In routine practice, a hysterosalpingogram will always be required in cases of habitual abortion to exclude such conditions as neoplasm or congenital anomalies of uterine shape, size and position. At this stage, the opportunity should be taken to examine the uterine isthmus and cervical canal, by placing the rubber cone close to the tip of the cannula. This method will suffice for the diagnosis of many cases of cervical incompetence. However, should no causative factor be demonstrated, or should the assessment of cervical incompetence be equivocal, the more accurate examination of the cervix by a balloon catheter method is indicated. For further reading see papers by Youssef (1958), Mann (1959), and Reiss (1959).

Complications

Many of the complications of hysterosalpingography may be deduced from the remarks made when discussing the contra-indications to this procedure.

1. **Abdominal pain.**—Central lower abdominal discomfort is often experienced from the instrumentation, but this is not usually severe and passes off quickly after the instruments have been withdrawn. Occasionally, however, it is sufficiently severe to mimic an 'acute abdomen' and may necessitate admission for observation. In the presence of cornual occlusion, either organic or spastic, persistence with the injection may induce central lower abdominal pain, due to the increase in intra-uterine tension. If the site of occlusion is in the tubes themselves, the discomfort induced will be felt near to or away from the mid-line according to whether the occlusion is in the proximal or distal part of the tubes. Peritoneal spill may induce discomfort to a varying degree. In this respect Neo-hydriol is superior to water-soluble media, but the occurrence of pain due to peritoneal spill seems to depend at least to some extent upon the pain threshold of the individual. In some patients the pain threshold is very low, whereas in others examined under exactly the same conditions no discomfort is felt. The pain usually passes off fairly quickly but may last for several hours, fading gradually. The patient should be permitted to lie down until the pain has disappeared, and should always be seen by the radiologist before leaving the X-ray department. Delayed abdominal pain occurring sometimes many hours after the examination should be regarded more seriously, particularly if it is accompanied by an elevation of temperature. In most cases no cause for the delayed symptoms will be discovered, and the symptoms subside with pethidine. Occasionally pelvic inflammation may be suspected necessitating antibiotic treatment. Discomfort at the time of the initial examination can be reduced as much as possible by ensuring that any traction applied to the tenacula is carefully controlled, by avoiding injection of contrast medium at excessive pressure and by stopping the injection as soon as peritoneal spill occurs.

2. **Perforation of the uterus or cervix.**—These complications are very unlikely if adequate precautions are observed. The preliminary passage of a uterine sound to determine the direction of the uterine cavity is important in the prevention of uterine damage, and the placing of the rubber cone fairly near the tip of the injection cannula limits the extent to which it can project into the uterine cavity. Bleeding after the examination should be regarded with suspicion, and admission considered.

3. **Ascending infection.**—This complication is uncommon when proper aseptic precautions are observed. As already indicated, symptoms suggestive of infection do not usually appear for some time after the examination. A hysterosalpingogram could, of course, light up an existing pelvic infec-

tion. Therefore, very careful consideration must be given to the advisability of proceeding with the examination when there is a history of recent pelvic inflammation.

4. **Abortion of an early pregnancy.**—Every so often a hysterosalpingogram is performed inadvertently in the presence of an early pregnancy. However, this diagnosis can usually be made after only a few ml. have been injected, and the examination is then stopped immediately. In actual practice, under these circumstances, abortion is unlikely but the patient should nevertheless be admitted for observation.

5. **Rupture of an ectopic pregnancy.**—As this possibility cannot be disregarded, a hysterosalpingogram in cases of suspected ectopic pregnancy should only be carried out in centres where operative facilities are readily available. It is unlikely that a hysterosalpingogram will be requested in acute cases. In these circumstances an emergency laparotomy is more usual, or, if less acute, a pelvic arteriogram may give the diagnosis without the danger of tubal rupture. In the more doubtful non-acute cases a hysterosalpingogram may show signs which are diagnostic, but often the examination is inconclusive.

6. **Venous or lymphatic intravasation.**—Reference to this complication has already been made when discussing contrast media. The factors which predispose to this are:—

1. Injection near the time of menstruation.
2. Injection of contrast medium at excessive pressure.
3. Direct trauma to the uterine mucosa by instrumentation.
4. Hysterosalpingography shortly after curettage or abortion.
5. Weakening of the uterine mucosa by conditions such as fibroids or tuberculosis.

If intravasation occurs when using oily media it is essential to observe the patient for some hours, preferably in hospital, although in practice pulmonary embolism is uncommon. Intravasation has not the same significance when water-soluble media are used as there is no danger of embolism.

With a little experience, the technique of hysterosalpingography is not difficult, particularly using the apparatus and technique described. It is particularly valuable to observe and control the whole procedure by screening. This will facilitate diagnosis and will give an indication as to the most suitable positioning of the patient for the purpose of radiographs. Screening is also helpful in evaluating the very common problem of uterotubal spasm by noting the variation in contour, whilst severe uterine spasm (ghost uterus) can only be appreciated fully by screen observation. Various complementary diagnostic radiological procedures can be combined with hysterosalpingography. Pelvigraphy is a natural further stage of hysterosalpingography, as already described. Pneumoperitoneum or pelvic arteriography may be successfully combined with hysterosalpingography, and can be very useful in the assessment of pelvic masses.

References

BARNETT, E. (1955). *J. Fac. Radiol.*, **7**, 115.

BARNETT, E. (1956). *Ibid.*, **7**, 184.

BROWN, W. E., JENNING, A. F., and BRADBURY, J. T. (1949). *Am. J. Obst. Gyneo.*, **58**, 1041.

CZYZEWSKI, W. J. J. (1956). *Br. J. Radiol.*, **29**, 679.

DAVIES, D. K. L., FISHER, H. J., and ROCHER, I. (1957). *Br. med. J.*, ii, 859.

HENRY, G. W., HUNTER, R. G. (1960). *Am. J. Roent.*, **84**, 924.

MCCANN, P., MENZIES, D. N. (1957). *J. Obst. Gynae. Brit. Emp.*, **64**, 416.

MANN, E. C. (1959). *Am. J. Obstet. Gyneo.*, **77**, 706.

NEMENOV, A. (1909). *Staatsverlag. Bd.*, **2**, 597.

REISS, H. E. (1959). *J. Obstet. Gynae. Brit. Emp.*, **66**, 317.

REISS, H. E., GROSSMAN, M. E. (1958). *J. Obstet. Gynae. Brit. Emp.*, **65**, 782.

SHEACH, J. M. (1959). *J. Fac. Radiol.*, **10**, 103.

WEISMAN, A. I. (1952). *Fertil. Steril.*, **3**, 290.

WRIGHT, J. T. (1961). *Br. J. Radiol.*, **34**, 465.

YOUSSEF, A. F. (1958). *Am. J. Obstet. Gyneo.*, **75**, 1305.

YOUSSEF, A. F. (1958). *Ibid.*, **75**, 1320.

APPENDIX

Trolley setting

Sterile—upper shelf

Injection syringe and cannula (the Green-Armytage type (Down Bros.) is recommended).

2 Volsellum forceps.

Uterine sounds.

Sponge forceps.

Towel (preferably with cutaway).

Gauze and wool swabs.

Gallipot with concentrated Dettol.

Bowl with dilute Dettol.

Speculum (or insertions of cold-light speculum—disinfected by immersion in hibitane)

Rubber gloves.

Lower shelf

Ampoules of contrast medium in a bowl of warm water.

Ampoules of pethidine, Phenergan.

Crushable ampoules of octyl nitrite.

Protective box.

Masks.

If a cold-light speculum is not used a suitable light will be needed.

The M-W cannula (now marketed as the VUC—Vacuum-Uterine-Cannula) is obtainable from Down Bros., Church Path, Mitcham, Surrey.

CHAPTER 12

ORAL CHOLECYSTOGRAPHY
(Prepared with the assistance of Dr. C. G. Whiteside)

CHOLECYSTOGRAPHY is so much an everyday examination that it is liable to be taken for granted. In fact, however, careful technique is required to produce films on which one can regularly give a confident opinion of 'no abnormality in the gall-bladder'.

Contra-indications.—The main contra-indication is the presence of jaundice and particularly obstructive jaundice. This is not an absolute contra-indication, for with levels of serum bilirubin below 5 mg. %, provided the cystic duct is not obstructed, cholecystography can be successful, especially when a double dose of contrast medium is administered. Even so it is sensible to improve the likelihood of obtaining good pictures by waiting until jaundice has subsided. *If cholecystography is attempted in the jaundiced patient, a high fluid intake is advisable to reduce the risk of renal damage.*

Preparation.—1. *Bowels.*—Routine bowel clearance is not normally used, but for repeat examinations in cases where the gall-bladder has been masked by the colon the patient should take an aperient, e.g. Dulcolax 1–2 tabs. for two nights before the examination.

2. The evening before the examination a light fat-free meal is taken. No food is taken on the morning of the examination. Any fluid taken should be free of milk to avoid inducing premature gall-bladder contraction.

Contrast media.—For straightforward oral cholecystography a number of media are available, any of which will opacify the normally functioning gall-bladder. Iopanoic acid (Telepaque) is the preparation most widely used at the moment but several other products are on the market and comparison of such media is the subject of a number of papers (e.g. Telepaque and phenobutiodil (Bryce and Cant, 1957), Pheniodol, Telepaque and Biligrafin (Gray, 1958), For the purposes of this chapter Telepaque will be discussed as the representative of this type of cholecystographic medium.

One drawback of simple cholecystography is that the bile-ducts are not regularly demonstrated even on the after-fatty-meal film. This is a significant defect in the procedure for two reasons:—

(i) Because the common bile-duct is an important part of the biliary tree and the presence of stones in the bile-duct should be demonstrated whenever possible. Furthermore, if the bile duct opacifies and the gall-bladder fails to do so there is a very strong inference that the cystic duct is obstructed and that therefore the gall-bladder is abnormal.

(ii) It is of considerable value to obtain a film which shows the bile-duct, even when it is normal in calibre. Such a film provides a 'base-line' for the size of the duct and if, following cholecystectomy, the duct is shown to have increased in calibre it can be assumed that it is obstructed. (The belief that a physiological dilatation of the bile duct occurs after cholecystectomy has been shown to be incorrect (Le Quesne, Whiteside and Hand, 1959)).

A contrast medium which goes some way to overcoming this difficulty is sodium or calcium ipodate (Biloptin or Solu-Biloptin; Schering). This substance is rapidly absorbed from the gut and is excreted by the liver in sufficient concentration to opacify the bile-ducts in some cases. To obtain a demonstration of the biliary tree the medium is given by the fractionated method—half the dose (3 G.)[1] being taken the evening preceding examination and half (3 G.) on the following morning 3 hours before examination. This product is now in routine use with satisfactory results (Whiteside, 1960, 1961; Murray, 1962).

Dosage of contrast.—For the average patient 3 G. of Telepaque are sufficient but a double dose (6 G.) is given:

(*a*) If the patient is over 12 stone (170 lb.) or unduly fat.

(*b*) If the initial examination with 3 G. gives poor delineation of the gall-bladder so that stones cannot be excluded.

Summary of the procedure

1. Prone film gall-bladder area when appointment made. If obvious gall-stones are present, change to intravenous cholangiogram.
2. Examination:
   (*a*) Prone film (prone-oblique, right side raised, in asthenic patients).
   (*b*) Supine-oblique film (left side raised).
   (*c*) Horizontal ray film.
   (*d*) If (*a*), (*b*), (*c*) satisfactory, give fatty meal.
   30–60 minutes later *coned* A.F.M. supine-oblique or prone.

FILMS

1. **Plain film.**—A plain prone film is taken at the time of booking the examination and the exposure is recorded on the request form. This film should include the right iliac crest, the spine, the right hemidiaphragm and the lateral abdominal wall. In cases which show undoubted stones it is not really necessary to carry out oral cholecystography; intravenous cholangiography (or at least oral cholangiography with Biloptin or Solu-Biloptin) is the more appropriate examination since the state of the bile-ducts and the relationship of opacities to them is information of more

---

[1] It is cheaper and equally satisfactory to give Telepagne (3G.) on the preceding night followed by Soln-Biloptin (3G.) on the morning of the examination.

value than to know how well the gall-bladder concentrates. This point is worth discussing with one's surgical colleagues. When no opaque stones are seen simple cholecystography may be carried out.

2. **Films at the examination.**—*Prone.*—The projection most commonly employed is the prone film. This position is adopted so as to bring the gall-bladder as close as possible to the film. In average or fat patients the flat prone position is satisfactory but in thin patients the gall-bladder often lies medially and a prone-oblique film is needed to bring it clear of the spine. (The likely position of the gall-bladder can often be assessed by looking for the liver edge on the plain film.) The prone-oblique position involves raising the right side, which takes the gall-bladder away from the film; the least possible degree of obliquity is therefore best.

Supine.—The supine-oblique (30° left side raised) position gives a good view of the neck of the gall-bladder and may produce some movement of stones in the gall-bladder. A combination of prone and supine-oblique films helps to reduce interference by gas in the hepatic flexure and displays differing aspects of the gall-bladder. Contrast is heavier than bile and the fundus is best shown in the prone position while the neck and body are best seen on the supine film. A modification of the supine view which may help to separate gall-bladder and colon is the film with the patient in the Trendelenburg position; this position is also of value for the low-lying gall-bladder. The supine-oblique view may be omitted in younger patients if the prone is satisfactory—in this case a supine-oblique after-fatty-meal film should be taken.

Film with horizontal ray.—At some stage in the examination a film with horizontal ray is taken. The aim of this is to show whether any filling defects visible on the other films are mobile and, in addition, to show small stones which may fall to the lowermost part of the gall-bladder or float in the contrast medium. The available methods are:—

A. *Screening with 'spot' films, patient erect.*—This, in our view, is the best way since films can be taken with varying rotation and compression so as to clear gas shadows from the gall-bladder and show the gall-bladder from different angles. It is not always easy to project the gall-bladder free of colonic gas; one manoeuvre which can be of value is to rotate the patient *slightly* to the left and then take a film in expiration with firm compression. The gall-bladder can be brought just clear of the spine and shown from a different projection. Although the preferred method, the use of fluoroscopy means that cholecystograms must be carried out when a radiologist is available to screen the patient and this is not always possible.

B. *Erect film.*—This is taken on a vertical Bucky, P.A. or P.A. oblique according to the patient's build. In centring, allowance is made for some descent of the gall-bladder. The drawback is that gas shadows sometimes overlie the gall-bladder.

C. *Lateral decubitus film.*—In this method the patient lies on the right side on some form of padding, preferably thick plastic foam. The film may be taken on a vertical Bucky, when the patient lies on a trolley; it may also be taken using a stationary grid with the patient on the X-ray couch. Centring is at the tip of the lowest palpable rib (i.e. 11th or 12th rib) and the film is taken P.A. with horizontal ray. The advantage of this view is that the gall-bladder and hepatic flexure tend to move in opposite directions.

3. **Other techniques.**—Sometimes, to show the gall-bladder clear of gas, films supine with 25° cephalad angulation or prone with 25° caudad angulation may be used. Tomography in the prone or prone-oblique position (3–8 cm. from the table top) is occasionally needed but we have very rarely found it as useful as good films taken with horizontal ray.

4. **After fatty meal** (A.F.M.).—When the initial films have been shown to be satisfactory the fat meal is given. This takes many forms; the simplest is Prosparol 30–60 ml.: for those who find the usual fatty meal nauseating a bar of milk chocolate is an adequate substitute. 30–60 minutes later a prone-oblique or supine oblique film is taken. As mentioned earlier the prone film shows the gall-bladder itself most clearly while the supine-oblique is commonly more suitable for showing the cystic and bile-ducts. It is usually best to choose whichever has shown least interference by gas earlier in the examination. Films taken during the first half-hour after the fat meal are more likely to show the bile-ducts but to demonstrate a well-contracted gall-bladder it is necessary to take a film 45 or 60 minutes after the fatty meal. If a sufficient number of films is taken—e.g. every 10 minutes, starting 15 minutes after fatty meal—it is usually possible to be certain of showing the bile-ducts; this is seldom convenient in routine practice.

General comments

1. *Kilovoltage.*—The kilovoltage should be low, in the region of 55–65 if possible; this applies to films taken while screening as well as to over-couch films. At the same time exposures should be kept as short as they can be—if necessary by shortening the focus-film distance.

2. *Immobilisation.*—A Bucky band is utilised to keep the patient still and, when used in conjunction with compression pads applied to the right hypochondrium, will help further to improve definition. With obese patients and a limited tube output, hyperventilation apnoea (p. 3) may usefully be employed.

3. *Coned views.*—Centring of films is facilitated by considering the patient's build and by relating the liver edge, seen on the plain film, to the iliac crest. However, the initial films can seldom be closely coned. Once the position of the gall-bladder is known, coning down is used to give a sharper picture.

Special techniques—the Four-day Telepaque test

It is sometimes possible to opacify stones in the biliary tree by prolonged administration of Telepaque (Salzman *et al.*, 1959). This applies mainly to stones in the bile-duct under conditions of stasis and mild jaundice. The medium reacts with the biliverdin on the surface of the stones, opacifying them. On occasion the bile-duct may also be opacified in this way. The indications are:—

i. Post-cholecystectomy cases showing a dilated bile-duct on intra-venous cholangiogram but no evident stone.

ii. Failed intravenous cholangiogram with mild jaundice and a clinical history suggestive of stone.

3 G. Telepaque are taken daily—1 G. three times a day after meals—for four days. A plentiful fluid intake is encouraged. On the morning of the fifth day, after a light fat-free breakfast, the gall-bladder area is X-rayed (prone or supine-oblique).

It should be noted that failure to produce an opacity does *not* exclude a calculus in the duct since some stones do not opacify.

Difficulties of Cholecystography

1. **Failure of opacification** or limited opacification of the gall-bladder. By its nature this examination allows a number of possible interruptions to the medium in its path to the gall-bladder and these possibilities should be reviewed when a gall-bladder appears not to function. The patient may fail to take the tablets or take an insufficient number or take them at the wrong time. Lesions such as pharyngeal pouch, achalasia, hiatus hernia or pyloric stenosis may delay the medium or diarrhoea may carry it too rapidly through the bowel. A fat-containing breakfast may have emptied the gall-bladder before the examination.

A repeat examination with a double-dose may therefore be carried out in any case where the 'non-function' is thought to be spurious or where, because of poor opacification, the examination is inconclusive. If diarrhoea due to the medium is thought to have impaired absorption, a different medium is given.

When the gall-bladder is thought to be truly non-functioning an intra-venous cholangiogram is carried out, *after a few days* as mentioned in Chapter 13.

2. **Gas obscuring the gall-bladder.**—This can usually be overcome but if it is impossible to obtain films with the gall-bladder clear of the colon an enema may be given or the examination repeated after taking an aperient for 2 nights previously.

3. **Reactions to the contrast medium.**—The commonest are nausea, vomiting and diarrhoea, occasionally of sufficient severity to interfere with absorption of the medium. Other, less common effects are abdominal pain, headache, facial flushing, urticaria or bizarre skin sensations. Urticaria is treated with antihistamines but otherwise treatment is symptomatic only. Serious reactions are very rare indeed.

References

BRYCE, A., and CANT, R. F. (1957). *Br. J. Radiol.*, **30**, 382.
GRAY, E. D. (1958). *Proc. R. Soc. Med.*, **51**, 793.
LE QUESNE, L. P., WHITESIDE, C. G., and HAND, B. H. (1959). *Br. med. J.*, i, 329.
MURRAY, J. P. (1962). *Br. J. Radiol.*, **35**, 278.
PHILP, T. (1961). *Br. J. Radiol.*, **34**, 58.
SALZMAN, E., SPURCK, R. P., KIER, L. C., and WATKINS, D. H. (1959). *J. Am. med. Ass.*, **169**, 334.
WHITESIDE, C. G. (1960). *Br. J. Radiol.*, **33**, 124.
WHITESIDE, C. G. (1961). *Ibid.*, **34**, 295.

APPENDIX

1. The instructions for preparation before a cholecystogram are supplied with every packet of tablets and will not be repeated here. For a double dose cholecystogram the instructions given to the patient are:—

(Double Dose) Cholecystography

On....................
At 1 p.m. have a normal lunch (preferably with an egg).
At 4 p.m. tea (without milk) and plain biscuits.
At 5 p.m. take one packet of tablets.
At 7 p.m. have a light supper. Drink plentifully.
At 10 p.m. take the second packet of tablets.
Have nothing further to eat or drink other than a cup of water or tea WITHOUT milk.
Come for your examination at................on....................

2. Prosparol (Duncan, Flockhart and Co., Ltd.) is an oil-in-water emulsion containing 50% of edible oil.
Dose 30–60 ml. (1–2 fluid ounces).

CHAPTER 13

INTRAVENOUS CHOLANGIOGRAPHY

(Prepared with the assistance of Dr. C. G. Whiteside)

THIS examination is indicated when the biliary tree is to be demonstrated in a patient who has undergone previous cholecystectomy or in one whose gall-bladder does not opacify with oral media. The density of the contrast medium (Biligrafin) when excreted and concentrated by the liver is normally sufficient to opacify the gall-bladder without further concentration by the mucosa. It is important to understand this difference from the conventional oral cholecystogram and to appreciate that a gall-bladder which has been demonstrated by this means may nevertheless be diseased.[1]

Contra-indications

1. *Jaundice.*—The conventional examination is generally found to be valueless with a serum bilirubin of more than 3·5–4·0 mg. % if the jaundice is fading and 1·5–2·0 mg. % when the jaundice is increasing. It now appears that the drip infusion method (Allen, 1969) may be successful in some cases (see below). Alternatively in these circumstances the four-day Telepaque test may be of value (p. 96).

2. *Allergy.*—This problem is similar to that encountered with urography and the matter is discussed under that heading in Chapter 3.

3. *Recent cholecystography.*—There is some evidence (Finby and Blasberg, 1964; Ansell, 1968) that intravenous cholangiography carried out shortly after oral cholecystography carries an increased risk of producing severe reactions. Moreover the concentration of contrast in the bile duct may be impaired. Thus, although the 'follow-on' intravenous cholangiogram is convenient, it is probably better to leave an interval between the two examinations.

Preparation

A. *Bowels.*—An aperient such as cascara tabs. 2 or Dulcolax tabs. 1–2 is given for two nights beforehand.

B. *Breakfast.*—The patient has a light fat-free breakfast on the morning of the examination.

[1] Although Biligrafin will opacify the non-concentrating gall bladder in some cases this does not justify its use as the initial means of investigation in all cases of suspected gall-bladder disease. There are several disadvantages to using the method as a routine—it requires an intravenous injection, the examination takes considerably longer and 'layering' often interferes with delineation of the gall-bladder. Moreover, Biloptin will often opacify the non-concentrating gall-bladder and has rendered the use of Biligrafin for this purpose largely unnecessary. Oral cholecystography thus remains the first choice in the investigation of gall-bladder disease.

C. *Fluid intake.*—To reduce the risk of renal damage patients should be encouraged to drink plentifully; this applies particularly to patients with renal failure.

Booking.—When booking make sure the X-ray table to be used is one suitable for tomography.

Preliminary film.—A prone-oblique (right side raised) or supine-oblique (left side raised) film is taken to show any opacities or translucencies related to the biliary tree and to give the correct exposure. The centring point is marked on the skin. At the time when this film is taken the patient should be asked about any previous allergy.

Contrast.—The contrast medium in common use in the United Kingdom in the past 15 years has been Biligrafin (Iodipamide–Schering). Recently a similar contrast medium, Bilivistan (Ioglycamide–Schering) has been introduced and it is claimed to have fewer side-effects and to show more prolonged opacification of the duct (Bell and Braband, 1966; Govoni and Toti, 1969). We have no personal experience with the latter medium but it seems possible that it will come into increased use in the next few years.

Administration of contrast

1. *Conventional method.*—40 ml. of Biligrafin forte (50% Iodipamide) or Bilivistan (Ioglycamide) are injected into a vein. An initial 1 ml. is given first and the patient observed for a minute. The full injection is then given over a period of about 3 minutes. Reactions are more frequent than during intravenous urography, nausea and vomiting being the commonest. The media are relatively viscid and it is advisable to employ a fairly large needle, e.g. Gillette 19G–1 R or S–TW.

2. *Drip infusion method.*—This technique (Allen, 1969) has been suggested as a means of improving biliary concentration of contrast in patients with jaundice or in whom the conventional method is unsuccessful. A rather larger dose of contrast (e.g. 60 ml. Biligrafin forte or 50 ml. Bilivistan) is given diluted in 200 or 250 ml. of up to 20% dextrose[1] and is run in slowly, taking an hour. Doses of up to 1 ml./kg. of Biligrafin forte can be given with improvement in duct opacification. This method allows longer for the contrast to become protein-bound. It is also possible by using larger volumes of fluid, to reduce further the risk of renal damage.

Summary of the procedure—conventional method

1. Plain film—enquire about allergy.
2. Slow injection of 40 ml. Biligrafin forte after test injection of $\frac{1}{2}$–1 ml.
3. Supine-oblique films of G.B. area at 15, 30, 45, 60 minutes after injection—later if necessary, and 90 and 120 minutes when the gall-bladder is present. It may be necessary to wait 4–5 hours for adequate

---

[1] The necessity for a high concentration of dextrose is as yet undecided; the evidence is conflicting. In most cases adequate results are obtained with saline or 5% dextrose.

gall-bladder filling. Tomography of the ducts as soon as they are visualised.

4. When the gall-bladder has filled, proceed as for cholecystography.

Position.—This may be supine-oblique or prone-oblique. Supine-oblique is pleasanter and easier to maintain for a period and may therefore be preferable when tomography is being carried out. The prone-oblique is, however, more widely used and probably produces a less rapid emptying of the bile-ducts, being comparable to the head-down position in urography.

Films.—As in oral cholecystography, a relatively low kilovoltage is employed. The films are taken at 15, 30, 45 and 60 minutes following the injection. The 15-minute film is usually a 'sighting shot' to help with correct centring. The ducts are ordinarily best seen between 30 and 45 minutes, so that films taken at these times should be well coned down to the biliary tree, having ascertained its position from the 15-minute film. Usually the ducts become fainter after 60 minutes but an obstructed duct may take longer to reach its maximum opacity particularly if it is dilated. In the absence of dilatation, persistent opacification of the duct after 60 minutes is unusual and may give an indication of the presence of partial obstruction (Wise *et al.*, 1957). A film at 2 hours should therefore be taken in post-cholecystectomy cases whenever there is doubt as to whether the duct is obstructed. (Note, however, that a negative result does not exclude partial obstruction.)

When the gall-bladder is present it will usually be apparent on films taken at 90 or 120 minutes but it may not appear until after 4 or 5 hours. In these circumstances the patient takes a light, fat-free lunch before returning for the late film. Once the gall-bladder has filled, erect and after-fatty-meal films may be taken as in cholecystography (p. 94). This is, of course, unnecessary if adequate films have already been obtained at cholecystography and when the examination is designed principally to show the bile-duct. Negative shadows or 'layer' shadows may be seen in the gall-bladder, particularly in erect or decubitus films; they are due to incomplete mixing of contrast and bile. These shadows are themselves confusing and may mask genuine stones. When they are seen an erect A.F.M. film should be obtained, as any spurious filling defects will have disappeared by the time this film is taken.

Tomography.—During an intravenous cholangiogram tomography is useful in the following circumstances:—

1. When gas or faeces obscure the duct.
2. When obesity makes the duct difficult to see.
3. When contrast is faint.
4. When the relationship of possible opaque stones to the duct is uncertain.
5. When the duct is dilated but no stone or causative lesion is seen.

6. If opacification of the right renal pelvis or the duodenum masks the lower end of the duct.

All this amounts to saying that unless the ducts are perfectly displayed they should be tomographed. The decision to do so should be made in good time so that preliminary cuts are taken by 30–45 minutes and the full 'run' within the hour. In the presence of delayed excretion tomography is deferred until the ducts are opacified throughout their length. It is helpful to take two preliminary cuts about 2 cm. apart, since one can judge from these which one is nearer to the actual plane of the ducts and thus plan the final run. In most patients in the supine-oblique position, 11 and 13 cm. are suitable levels; in the very fat, 13 and 15 cm. Corresponding levels for the prone-oblique position are—thin patients, 7 and 9 cm., average 8 and 10 cm., fat patients 10 and 12 cm. It is unusual to show the entire duct on one 'cut' and it is necessary to make sure that sufficient films are taken to show the duct 'in toto'.

Summary of the procedure—drip infusion method

1. Plain film—enquire about allergy.
2. Set up drip of 250–500 ml. Biligrafin-dextrose (or Bilivistan-dextrose.) Allow to run in over a period of an hour.
3. Prone-oblique film of bile-duct and gall-bladder followed by tomography.
4. Delayed films, if needed, to allow the gall-bladder to opacify.

Procedure

The contrast must be freshly mixed when using 20% dextrose since the composite solution caramelises during autoclaving. The drip is administered through a disposable giving set and a catheter needle—e.g. Plextrocan or Teflon Longdwel—and adjusted to run in over about an hour (approximately 100 drops per minute). This can, of course, take place outside the X-ray Department if necessary. When the infusion is complete the catheter is removed and the patient transferred to a radiographic room. A prone-oblique film of the gall-bladder area is followed by routine tomography as described earlier. The gall-bladder is usually filled at this stage but it may occasionally be necessary to take films at 3–4 hours to confirm that opacification is not occurring.

The indications for this method are not yet fully defined. Its advantages are the greatly reduced incidence of side-effects, the shortened time necessary for radiography and the apparent improvement in opacification of the duct system; it may well be that it will come to be regarded as the preferred method for routine work.

Special problems—acute cholangiography.—In patients who have acute abdominal symptoms suggestive of gall-bladder disease it is possible to provide a relatively rapid assessment of the biliary system comparable to acute urography in suspected renal colic. Following a plain film of the gall-bladder area, contrast is given by either of the methods described above, preferably by a drip. A film of the gall-bladder region is taken an hour later. If it shows normal outlining of the ducts and gall-bladder, biliary disease is virtually excluded. The presence of gall-bladder disease is almost certain if the gall-bladder fails to opacify by 4 hours (Weens and Walker, 1964).

Difficulties.—Some of the difficulties of the examination have already been mentioned—gas and faeces in the colon, 'layer' shadows in the gall-bladder, and masking of the bile-ducts by contrast in the duodenum or right renal pelvis and ureter.

Another source of difficulty arises when the contrast medium fills the duodenum and refluxes into the cap. This can then give a shadow similar to that of the gall-bladder. The diagnosis is usually obvious but it may be necessary to give a little barium or Gastrografin to confirm the position of the duodenum in cases where there is doubt as to whether cholecystectomy has been performed.

When the plain films show gas outlining the biliary tree, indicating a biliary-gut fistula, the normal examination is seldom successful since the translucency of the gas cancels the opaque contrast. Small quantities of gas offer problems of diagnosis rather than of technique since the ducts generally outline satisfactorily and the gas appears as translucent filling defects. The distinction from non-opaque stone is made (Samuel and Scott, 1958):

(i) from the plain film, which should also show the translucencies.
(ii) from the clear-cut spherical appearance of the translucencies which may often be multiple—when numerous duct stones are present they are usually facetted.

In cases of continued doubt, barium or Gastrografin may be given by mouth; if a fistula is present barium may then outline the biliary tree.

References

ALLEN, W. M. C., (1969). *Br. J. Radiol.*, **42**, 347.
ANSELL, G. (1968). *Clin. Radiol.*, **19**, 175.
BELL, H. D. C., and BRABAND, H. (1966). *Br. J. Radiol.*, **39**, 63.
FINBY, N., and BLASBERG, G. (1964). *Gastroenterology*, **46**, 276.
GOVONI, A. F., and TOTI, A. (1969). *Am. J. Roentg.*, **107**, 14.
SAMUEL, E., and SCOTT, W. (1958). *Br. J. Radiol.*, **31**, 631.
WISE, R. E., JOHNSTON, D. O., and SALZMAN, F. A. (1957). *Radiology*, **68**, 507.
WEENS, H. S., and WALKER, L. A. (1964). *Radiol. Clins. N. America*, **2**, 89.

APPENDIX

Instructions to prepare for intravenous cholangiogram:—

Your X-ray appointment time is at . . . . on . . . .
Take an aperient such as cascara, 2 tablets, for two nights before the examination.
No greens, cereals, potatoes or bread should be taken on the day before the examination but you should drink plentifully.
Take only a light breakfast on the morning of the examination.
The examination will last up to 5 hours.

Tray or Trolley

As for intravenous urography with 2×20 ml. ampoules of Biligrafin forte or Bilivistan instead of urographic contrast.

   or 250 ml. bottle of 20 % dextrose or 500 ml. of 5 % dextrose.

   3×20 ml. ampoules of Biligrafin forte or Bilivistan (Schering).

   1×20 ml. syringe and pentothal mixer (for adding contrast to dextrose).

   Disposable sterile giving set.

   Needle-cannula-Plextrocan (P.P.) *or* Longdwel (B.D.) 6710 or 6717.

CHAPTER 14

PERCUTANEOUS TRANSHEPATIC CHOLANGIOGRAPHY

(W. B. Young, F.R.C.S.Ed., F.F.R., D.M.R.D.)

PERCUTANEOUS hepatic cholangiography by the method to be described is carried out by introducing a polythene tube directly through the skin into one of the intrahepatic bile ducts and injecting radio-opaque fluid through it to opacify the duct system.

Indications

1. Jaundice—persistent, recurrent, or increasing in severity.
   (*a*) Obstructive jaundice in which the nature of the obstruction is not certain.
   (*b*) Post-operative jaundice when thought likely to be due either to stricture or non-opaque stones.
   (*c*) Jaundice of over one month's duration—to differentiate between obstructive and intrahepatic cholestatic jaundice (primary biliary cirrhosis, drug jaundice, etc.) when clinical and laboratory methods have failed to provide a conclusive answer. (A negative result is of almost as much significance as a positive one.)
2. To determine the cause of recurrent cholangitis.
3. To decompress or drain the bile ducts in cases of obstructive jaundice where operation is temporarily contra-indicated.
4. It has been suggested that it might be used as a means of injecting antibiotics directly into the biliary tree in cases of chronic infection.

Discussion

A number of different workers have described methods of injecting the intrahepatic bile ducts directly through the skin using a flexible steel needle. The majority have advocated an anterior sub-costal approach though some prefer a lateral inter-costal route. Others recommend peritoneoscopy to guide the needle under direct vision into the gall bladder, or between the layers of the falciform ligament into the liver to achieve an extra-peritoneal approach. A posterior approach through the bare area of the liver has been recommended by some for the same reason. None of these methods, except perhaps the last, has been free of occasional serious complications, the most frequent of which has been bile peritonitis, and a number of deaths have been reported: at least one death from haemorrhage due to liver trauma has also been described. As a result the procedure has fallen into disrepute and has been condemned as dangerous by some writers.

The method to be described was devised by Dr. Stanley Shaldon, Lecturer to the Medical Unit of the Royal Free Hospital, London, and over 350 examinations have been made in the X-ray Department without fatality or serious complication.

A fine polythene tube is introduced on a 20-gauge needle some 15 cm. long directly through the skin into the liver. The needle is at once removed leaving the polythene tube behind. This tube is gradually withdrawn while suction is applied to its free end, and once a duct is entered bile is aspirated into the tube where it can be recognised by its colour or consistency. Sufficient opaque medium is then injected under X-ray screen control to obtain optimum filling of the duct system, and films are taken in those positions thought best to show the site and extent of any obstruction present.

Special points.—The examination is carried out in the X-ray Department on a standard tilting screening table. Image intensification is an advantage. A general anaesthetic is not used except in the case of children or very nervous patients.

Ward preparation

1. The prothrombin time is checked and corrected if necessary.
2. Arrangements are made for laparotomy immediately following the procedure. Abdominal skin preparation is carried out in the ward and food and drink is withheld on the morning of the procedure. The patient is premedicated with Omnopon and scopolamine 45 minutes before the time booked for the X-ray examination.

Department Preparation

1. Preliminary preparation of tubes and needles.[1]—The tubing used is thin-walled flexible polythene PE 160. Short lengths are cut, each approximately 20 cm. long. A flange is made at one end of each of these short tubes by rotating the tip over a small flame to ensure a good fit inside a tap adaptor (p. 239). These tubes are then packed in small paper envelopes, each containing six, and sterilised by gamma radiation. Sets of six flexible 20-gauge steel needles, approximately 15 cm. long, ground with a Pitkin cutting point, and two tap adaptors, are packed in glass test tubes and autoclaved.
2. Preparation of equipment for the examination.—A sterile trolley is prepared—see Appendix.

[1] *Editor's note*: The experience of the Royal Free Hospital shows that polythene tubing is extremely safe and clearly Dr. Young would be unwilling to ratify the use of any other materials. It may, however, be pointed out that Teflon needle catheters have been successfully employed for this examination and that they are free from the tendency to buckle, which is the drawback of polythene.

Technique of examination

While the examination can be performed by a single operator it is best carried out by two, one of whom intubates the biliary tree while the other operates the fluorescent screen and takes films. It is carried out as a sterile procedure, the operator being gowned and gloved. All persons in the room wear masks and lead aprons.

The patient lies supine on a motor-driven tilting screening table. The skin is prepared with spirit, landmarks are identified, and sterile towels applied. The site for insertion of the needle is chosen approximately 3 cm. below and 3 cm. to the right of the xiphisternum. Local anaesthetic (1% lignocaine) is injected into the skin and then more widely into the soft tissues down to the liver capsule.

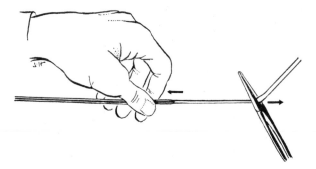

FIG. 11. Catheter preparation (see text).

A packet of tubes is then opened, together with a tube of needles and adaptors. A needle is slipped through the flanged end of one of the polythene tubes until the flange is about 1 cm. from its mount. Holding the tube and needle tightly in one hand, the free projecting end of the tube is grasped close to the needle point and pulled out and away from the needle, so that it is stretched and reduced in calibre. A firm pull is required (Fig. 11). The attenuated portion of the tube is then cut off flush with the tip of the needle and the tube is milked up the needle until the bevel projects just beyond the end of the tube, which should now fit snugly round the end of the needle. The end of the tube is inspected carefully for signs of splitting and if it is damaged in any way it is discarded. Six needles and tubes are prepared in this manner and laid on the sterile trolley ready for use.

A small incision is made in the skin through the wheal raised by the local anaesthetic, and the points of a pair of mosquito forceps are introduced and forced through the thickness of the subcutaneous tissue and rectus muscle using a rotating motion of the wrist. The forceps are opened from time to time as the points are advanced to stretch the tissues down as far as the peritoneum. It is essential to make a good track for the needle and tube in this manner, especially in patients whose abdominal wall is scarred from previous operations. One of the needles with its tube is slipped through the collar of a tap-adaptor (Fig. 12a) and inserted along the track made by the forceps, and a 20 ml. syringe half filled with 1% lignocaine is attached. The patient holds his breath in full expiration and the needle is slowly advanced into the liver substance being directed a little upwards and laterally, while an injection of anaesthetic is made. Once the needle is fully inserted to its hilt, it is withdrawn leaving the polythene tube in the liver. The patient may now breathe normally.

The two parts of the tap-adaptor are screwed together and a 20 ml. syringe containing about 10 ml. of normal saline is connected. With the

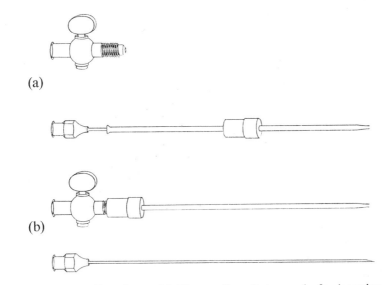

(a)

(b)

Fig. 12.—Needle-catheter. (a) The needle-catheter ready for insertion with the collar of the tap-adaptor in position; the butt of the adaptor is shown above. (b) The catheter assembled, with the needle below it.

tap open a little saline is injected to make sure the tube is clear. A gauze swab is inserted behind the tube to provide a white background, and while gentle suction is maintained on the syringe the tube is very slowly withdrawn from the liver. If a blood vessel is entered the tube immediately fills with blood. This is of no consequence and the tube is flushed with fresh saline and withdrawal continued. When a bile duct is entered there is an immediate change of pressure in the syringe and bile is aspirated into the tube. Normally it is easily recognised by its colour, but occasionally the bile is white and in this case it is always under pressure and there is a free flow of colourless fluid up the tube which continues when the syringe is disconnected. The escaping fluid is sticky to the touch. A few ml. of bile is drawn off and injected into a sterile bottle for culture, the tap is closed and the syringe disconnected. The tube is secured to the skin by means of a suture or by applying adhesive tape. A 20 ml. syringe is filled with opaque medium (we use 45% Hypaque) and attached to the adaptor by means of a polythene cistern, air bubbles having been first excluded. The tap is opened, lights are dimmed, and under X-ray screen or television control, a little contrast fluid is injected. The radiologist operating the screen can see immediately whether contrast is flowing into the ducts or not; if it is, the injection is continued until he gives the word to stop. The amount of opaque medium required varies from case to case and depends on the capacity of the duct system. 20 ml. may be sufficient but 60–100 ml. or more may be required if the duct system and gall-bladder are markedly distended. Hypaque seems to mingle readily with the bile and there is no need to draw off much bile before injecting.

The radiologist controls the amount injected and takes films at such times and in such positions as he considers to provide maximum information; he may rotate the patient or tilt the table in order to demonstrate the anatomy of the ducts as fully as possible. It may be necessary to take one or more films with the patient prone if the left intrahepatic duct system is to be properly shown. Using a polythene tube there is no danger of causing damage to the liver by moving the patient, or if he should cough. Three or four 12″ × 10″ films are taken in different positions of rotation and one or more series of spot films using a small compression cone.

After the films have been taken the tap is opened and the end of the extension tube is placed in a sterile bottle to drain the ducts.

The patient remains on the X-ray table until the films are developed and inspected. If, after examining them, it is thought necessary, more opaque medium can be injected and further films taken. If the films are satisfactory sterile dressings are applied over the puncture site and the patient is transported to the theatre with the polythene tube and extension still draining. When everything goes well the procedure takes about 20 min.

If the tube is completely withdrawn without bile being aspirated a second puncture is made using a fresh needle and tube—directing the needle a little to the right of the first puncture. If this fails another needle and tube is directed a little to the left. We do not make more than six punctures in all for it is our experience that if the duct system is dilated a duct is usually found at the first or second puncture. If the first four punctures are unsuccessful at times we make a lateral puncture through the 9th or 10th intercostal space in the mid axillary line depending on the position of the diaphragm as seen on screen examination. The technique is the same and the needle is directed horizontally. With a high duct obstruction usually only the right hepatic duct system is visualised by this route. If a duct is not entered in six attempts it is unlikely that the ducts are dilated, and improbable that the jaundice is due to extra-hepatic obstruction. If no duct is intubated there seems to be little risk of subsequent leakage of bile into the peritoneal cavity, and so far none of our patients has developed symptoms which could be related to this complication. Laparotomy is usually postponed and the patient is returned to the ward for further observation and assessment.

Difficulties and Complications[1]

1. The polythene tube may buckle as it is being inserted. This is most likely to occur when the puncture has to be made through scar tissue resulting from previous operations, and means that the track prepared for the needle with the forceps is inadequate. The mosquito forceps should be re-inserted and a more satisfactory track made.

2. The patient may complain of a little pain during the procedure, often referred to the right shoulder, but this is rarely severe, especially if a little local anaesthetic is injected while introducing the tube. If opaque medium leaks under the liver capsule, or into the peritoneal cavity the patient may complain of fairly severe upper abdominal pain. This has occurred in only a few of our cases, and in each passed off within ten to fifteen minutes. On one occasion it was considered severe enough to warrant a small dose of pethedine.

3. The tube may be inserted directly into the gall bladder. This is immediately recognisable on the screen. We continue to inject opaque medium to try and demonstrate the full anatomy of the duct system. If the ducts are not dilated and there is no obstruction to the flow into the duodenum it is probably safe to leave the tube in to drain the gall bladder for 6–8 hours and to cancel the operation. The patient should be watched carefully over the next forty-eight hours for evidence of leakage into the peritoneum. If Hypaque has been demonstrated outside the gall bladder it is probably safer to proceed with laparotomy, mop out the peritoneal cavity, and deal with any leaking point that may be found.

4. Occasionally the stomach or duodenum will be intubated. There is then a free flow of rather thin fluid which may be greenish yellow in colour. The true state of affairs is usually immediately obvious after injecting a few ml. of contrast material under screen control. The tube is removed, and another puncture is made into the liver using a fresh needle and tube. There seems to be little danger of peritoneal leak under these circumstances, and in three cases in which we have punctured the stomach or duodenum the puncture sites have not been obvious at subsequent laparotomy and there have been no complications.

5. We have had only one case of intraperitoneal bleeding, though on several occasions clots and a little free blood have been found in the peritoneal cavity at laparotomy. In this case the bleeding area was easily controlled at laparotomy, an advantage of the anterior approach. The puncture sites in the liver are always examined for bleeding or bile leakage at laparotomy.

6. The most serious complication seems to be biliary peritonitis. If the polythene tube is left in the ducts to drain them until the time of

[1] *Editor's note*: The presence of contrast in the urinary tract on any of the films indicates that there may be a communication between a bile-duct and a hepatic vein. Should this occur there is a strong likelihood that the patient will develop a septicaemia and this is a sign that the patient should be treated appropriately (D. H. Nelson—personal communication).

operation, and provided laparotomy takes place within six to eight hours, there seems to be little, if any, danger of this, for once the obstruction in the duct system is relieved at operation, and pressure reduced, there is little risk of leak along the needle track. Any spill that has occurred as a result of the examination should be mopped up.

We, therefore, recommend that this examination should be performed only when a surgeon is prepared to undertake laparotomy if an obstructive lesion is demonstrated; and that the patient should have been prepared for operation and on his way to the theatre. We believe that under these circumstances the procedure is safe, and one that may provide the surgeon with extremely valuable information, whereby he can plan his whole operative procedure. At times it may furnish him with information no other method can provide.

If for some unseen circumstance after demonstrating an obstruction in the ducts it is impossible to proceed straight to operation then the polythene tube should be left in place to drain the ducts for at least six to eight hours, and longer if possible, to enable the puncture track to be sealed off from the general peritoneal cavity and to allow an external fistula to develop if pressure in the ducts remains excessively high.

After return to the ward.—If no duct has been intubated the patient is returned to the ward and put on a half-hourly pulse for 12 hours to give notice of possible haemorrhage; and the patient is watched for leakage of bile and the development of peritonitis. So far we have not had any complications of this kind.

If laparotomy has been performed normal surgical after-care is instituted.

APPENDIX

Trolley setting

Sterile—upper shelf
Bottle of sterile normal saline with three-way tap for irrigating polythene tube.
3–4 dressing towels. Gown.
A syringe and needles for injection of local anaesthetic into the skin and down to the liver capsule.
Set of polythene needle-catheters as described in the text *or* Teflon needle-catheter—Becton-Dickinson, Longdwel: 01–0050 or 01–0052.
Supply of gauze swabs. Several pieces of sterile strapping.
A fine scalpel.
Two pairs of mosquito forceps.
Three 20 ml. syringes with drawing-up needles.
Two polythene cisterns about 18 inches long for connecting the syringe to the tube by way of adaptor.
A skin suture needle threaded with silk.
Sterile gloves. Paper hand towel.
Gallipots for spirit or antiseptic.

Lower shelf
1 % lignocaine.
Sterile saline.
Three or four 20 ml. ampoules of 45 % Hypaque are kept standing in hot water.
Masks.
Specimen bottle for culture of bile.

CHAPTER 15

OPERATIVE CHOLANGIOGRAPHY
(Prepared with the assistance of Dr. C. G. Whiteside)

THE value of this procedure has long been debated by surgeons but there is little doubt (Samuel, 1959a; Le Quesne, 1960) that properly carried out, it can greatly increase the accuracy of assessment of the bile-ducts at operation. It is, as Patey (1960) observes, a surgico-radiological technique and the best results are obtained when a radiologist is present in the theatre to supervise the procedure and to help in interpreting the films. Most radiologists find it difficult to be regularly available in this way, but it may still be reasonable to introduce the procedure if the surgeon is convinced of its value and if he is willing to learn the details of the technique. The radiologist will then be present during the initial examinations until he is assured that a satisfactory routine has been established and that the essentials of an adequate result are appreciated by the surgeon.

A distinction must be made between pre-exploratory and post-exploratory cholangiography. The former is carried out early in the operation and its main purpose is to exclude any calculi in the bile-duct; it will frequently obviate the need for exploration of the duct by showing it to be normal. On occasions it may show abnormalities other than calculi, e.g. low implantation of the cystic duct, strictures of the bile duct or deformities due to pancreatic tumours.

Post-exploratory cholangiography is carried out after the T-tube has been inserted into the common duct. Its purpose is to ensure that no stones have been overlooked and it is clearly far more valuable to make such a check during the operation than ten days later at post-operative T-tube cholangiography. Until recently the post-exploratory examination has been found unreliable because of the difficulty of excluding air bubbles. Now, however, the use of continual irrigation during insertion of the T-tube has been incorporated in the technique (Whiteside, 1961) and this will usually result in a satisfactory examination.

Summary of the Procedure

Pre-exploratory cholangiogram

1. Position the patient on the operating table or tunnel with the gall-bladder area over the space to be used for radiography.
2. Centre the X-ray tube, then swing it out of the way.
3. When the cystic artery has been ligated, the surgeon exposes the cystic duct and catheterises it with a bubble-free saline-filled system.

(A catheter should not be passed beyond the lower end of the cystic duct.)

4. After aspirating bile the surgeon attaches a syringe of contrast medium and draws the plunger back again to catch the bubble included during attachment.
5. Tilt the table 10°–15° to the right.
6. The surgeon injects contrast slowly, stopping for three exposures.
7. He proceeds with cholecystectomy while the films are being developed.

Post-exploratory cholangiogram

1. Attach a 50 ml. syringe of saline to the T-tube and flush it through.
2. Insert the T-tube in the common bile-duct with continued slow irrigation of saline.
3. When the T-tube is well sewn in, the force of injection is increased to test for leaks.
4. Change to 20 ml. syringe of contrast, withdrawing the plunger *slightly* to catch any air included during attachment.
5. Inject contrast slowly, pausing for the exposures.
6. Expose films after 6, 12 and 18 ml. of contrast have been injected.

PROCEDURE

A. Radiographic equipment

1. Some form of tunnel is required, allowing cassettes to be slid under the patient. The apparatus described by Samuel (1959b) is excellent, enabling up to six films to be taken in succession; its especial advantage is that films are fed in from the end of the table so that the surgeon is not disturbed. In some operating tables there is a space into which cassettes may be placed. Bandages looped round the cassettes then enable them to be pulled out and changed.

2. A grid is needed and should be set *with lines running across the table*, thus allowing the table to be tilted sideways.

3. A sterile cover is fastened over the tube of the X-ray machine.

B. Contrast medium

Diodone 35% or Hypaque 25% are employed.

C. Technique—Pre-exploratory cholangiogram

The patient lies on the table with the gall bladder area over the film space. The portable machine may be positioned so that the tube merely has to be swung over the patient. When the cystic artery has been tied the cystic duct is cannulated; various types of tubing and cannula have been recommended for this including:—

(*a*) a ureteric catheter.

(b) polythene catheter PE 160 or PE 205 with a tap/adaptor as used in aortography.

(c) Stoke-on-Trent cholangiography cannula (PP).

(d) Hamilton-Bailey cannula on a plastic connecting tubing.

Our own preference is for polythene tubing because bubbles show easily. Where a cholecystectomy has already been performed, a hypodermic No. 20 needle on a plastic connecting tubing may be inserted into the bile-duct, the needle being positioned to one side in order not to overlie the duct. Whatever system is employed the following points should be noted:—

1. The system must be long enough to keep the operator's hands well clear of the field of irradiation.
2. The catheter system is filled with saline and free of bubbles. It is an advantage if a saline syringe is attached and an assistant slowly injects saline during cannulation.
3. The catheter is advanced to the common bile-duct—if it goes too far contrast may be delivered straight through the sphincter of Oddi.
4. Once the catheter has been tied in, aspiration is carried out until the bile appears in the saline syringe. Saline is then injected to test for leaks (syringe held with nozzle down).

Now a 20 ml. syringe of contrast medium replaces that of saline. Contrast may be Hypaque 25%, or diodone 35%. If there is a tap on the catheter it is turned off while the syringe is changed. The procedure is then as follows:—

1. Once again gentle aspiration is carried out to draw back into the syringe the air bubble which is almost always included during attachment.
2. The table is tilted 10°–15° to the right and the radiographer centres the X-ray tube to the surgeon's finger, placed on the entry of the cystic duct into the common bile-duct. All instruments, towel clips, stomach tubes, etc., are removed from the field of examination.
3. The anaesthetist is warned to be ready to stop the patient's respirations and the assistant who is to change the cassettes takes up a position, wearing a lead apron. The surgeon should be protected; a convenient mobile form of protection consists of a double-sided lead apron hung by a coat hanger on the lower part of a drip stand and covered in a sterile towel. The remaining assistants move away.
4. Injection is carried out slowly with the syringe angled downwards to reduce the risk of injecting bubbles. With a fine catheter films are taken when the following amounts have been injected:—

1st film—after 2–3 ml.
2nd film—after 7–8 ml.
3rd film—after 12 ml.

These amounts are increased if the connecting system is wide or if the duct is dilated. For each exposure the surgeon says 'now' to the anaesthetist and slows his injection. The anaesthetist arrests respiration and calls 'take' to the radiographer. As the surgeon sees that breathing has stopped he stops injecting altogether since the flow of contrast can make stones move and blur them. After each exposure, respiration and injection are resumed. Even when large amounts are injected, the last 1 ml. is kept in the syringe.

When three cassettes have been exposed the X-ray tube is removed and the operation is resumed. While the films are being developed—preferably close at hand—the cholecystectomy is completed. The results of the examination will determine whether the surgeon should proceed with exploration of the common duct or whether he can now pronounce the duct normal and close the abdomen.

Difficulties

1. *Air-bubbles.*—These can usually be recognised by their smooth outline and spherical shape. It is not always possible to distinguish air-bubbles from stones with certainty and it may therefore be necessary to flush the ducts with saline and repeat the examination.

2. *Spasm of the sphincter of Oddi.*—This may be suspected (Le Quesne, 1960) when the contrast medium fails to enter the duodenum easily although the duct is of normal calibre and no obstructing lesion is seen. Such spasm may be due to a small concealed stone or to the drugs used in premedication. Amyl nitrite introduced into the anaesthetic circuit will relieve spasm due to drugs and a repeat examination will show normal appearances.

Technique—Post-exploratory cholangiography

When stones have been removed from the common bile-duct it is extremely useful to be able to check that there are none left. Cholangiography through the T-tube is of value only if all air bubbles have been cleared from the duct and the T-tube. The technique is similar to that described above but with the following differences:—

1. When the T-tube is ready for insertion into the common duct it is connected to a 50 ml. syringe of saline. All air is flushed out. As the T-tube is inserted and while it is being sewn in, a continual slow irrigation is carried on, the assistant sucking away the excess saline.

2. When the T-tube has been sewn in the duct, the rate of injection is increased to test for leaks and if the result is satisfactory the 50 ml. syringe is exchanged for a 20 ml. syringe of contrast. A very gentle aspiration is then carried out to draw back any bubbles of air included during the exchange. Contrast is then injected, exposures being called as before, allowing about 4–5 ml. extra for the wide lumen of the T-tube.

N.B.—i. It is important not to attempt more than the gentlest aspiration at any stage since air can be drawn thereby into the duct.

ii. It is common for oedema of the sphincter of Oddi to prevent contrast from flowing into the duodenum as it should on a pre-exploratory cholangiogram. This obstruction will not be relieved by amyl nitrite, but provided the lower end of the common duct in these circumstances appears convex or tapers to a point, such obstruction can be ignored. A residual stone causing obstruction will be shown by a concave margin at the lower end of the duct.

In conclusion we would point out that accurate interpretation is all-important in making this a truly reliable investigation; the reader is advised to consult Le Quesne's (1960) or Norman's (1951) papers for guidance on the diagnostic aspects.

References

Operative cholangiography
LE QUESNE, L. P. (1960). *Proc. R. Soc. Med.*, **53**, 852.
NORMAN, O. (1951). *Acta Radiol.* Suppl. 84.
PATEY, D. H. (1960). *Ibid.*, **53**, 851.
SAMUEL, E. (1959a). *Br. J. Radiol.*, **32**, 669.
WHITESIDE, C. G. (1961). Lecture delivered at Autumn Post-graduate Week-end of the Faculty of Radiologists.

Serial tunnel
SAMUEL, E. (1959b). *Lancet*, i, 454.

CHAPTER 16

T-TUBE CHOLANGIOGRAPHY

THE aim of this examination is to outline the hepatic and common bile-ducts after cholecystectomy, to show them free from stones or narrowing and to demonstrate a ready flow of contrast into the duodenum. As with operative cholangiograms the technique must be faultless: if stones are shown they must be shown convincingly; equally the normality of a biliary tree must be indisputable. Normally the cross-limb (which is sometimes not a tube but a V-section channel) lies in the bile-duct but occasionally the T-tube has ceased to drain and the examination is then undertaken to show where it is lying.

Summary of the procedure

1. Position the patient, lying about 10°–15° onto the right side.
2. Clean T-tube, clamp it near to patient but away from field of irradiation.
3. Attach needle to a flexible connection and so to a syringe of 25% Hypaque; thrust through the wall of T-tube above the clamp. Inject slowly and take films under screen control.

Technique

1. The examination is best carried out under screen control not only because multiple spot films are helpful but because varying quantities of contrast medium may be required, depending on the condition of the bile-ducts.

2. The patient should come to the X-ray department with the tube spigoted. When she lies on the couch any removable dressings and safety pins are taken off. The left shoulder and buttock are supported with pads to incline her about 10°–15° on to the right side. Subsequently slight alterations are made under screen control to ensure that the duct is just off the spine. Too much rotation tends to throw the lower end of the duct over the contrast in the duodenum.

3. The operator takes hold of the tubing in a spirit-swab, wiping it with another spirit-swab. It is then laid on a sterile towel, drawing it down so that there are no redundant loops. It is clamped above any visible bubbles and again swabbed with spirit above the clamp.

4. 20 ml. of 25% Hypaque are drawn up, the syringe is attached to a flexible connection, and the latter is fixed to an intramuscular needle. *All bubbles are carefully expelled.* The needle is thrust through the wall into the

lumen of the tubing above the clamp (Fig. 13). The purpose of the connection is to keep the operator's hands well away from the field of irradiation. With long T-tubes it may not be needed.

5. Slow injection is commenced under screen control; the contrast is watched as it runs into the bile-duct, and 2–4 local exposures made as the bile-duct fills. Early films are important to determine the ease of flow into the duodenum and to show the narrow intramural part of the bile-duct before it is obscured by contrast in the duodenum. When these have been obtained an oblique 12″ × 10″ film of the whole biliary tree is taken. The

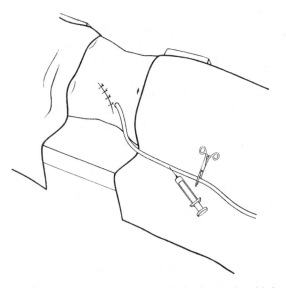

FIG. 13.—T-tube cholangiography. The tube is clamped and injection of contrast medium is made above the clamp. Note that patient lies slightly onto the right side. For the sake of simplicity the flexible connection mentioned in the text has been omitted.

hepatic ducts should be filled for this film and to achieve this it may be necessary to tilt the head of the couch down and inject forcefully with the patient taking deep breaths. The pads are then removed so that the patient can lie flat and a further 12″ × 10″ film is taken. It is helpful if the patient is brought into the semi-erect position for this film since any translucencies due to air-bubbles will be shown to move upwards (Dr. J. R. Barnett—personal communication).

All films are taken immediately after injection but injection is stopped for the actual exposure—as the operator calls 'stop breathing' to the patient he stops injecting.

The films are inspected carefully and if the appearances are conclusive the tube is unclamped and the patient sent back to the ward. Sometimes translucencies are seen which might be due to air-bubbles and then the following procedure is used:—

E

The patient is turned more than half-way onto the right side and tilted head down. 50 ml. of saline are injected using the same technique as before. Injection should be as forceful as the patient's discomfort will allow. Towards the end of the injection the patient is turned onto the back. The cholangiogram is then repeated.

Overcouch technique.—When screening is not possible overcouch films are taken after injection in the same way. Exposures are made after 6, 12 and 18 ml. have been injected.

Special problems

1. *Air-bubbles*. These are dealt with above.

2. *Obstruction at the lower end of the duct*. Flow of contrast into the duodenum is part of the normal examination. If an obstruction is found injection should be continued to the point of mild discomfort. Should the obstruction persist screening is stopped for 2–3 minutes and any films taken are sent for developing. Usually on screening again some contrast will be seen to have escaped into the duodenum but further small injections and a pause of up to 10 or 15 minutes may be necessary. Films must always be taken to demonstrate the obstructing lesion.

3. *The tube has pulled partly or wholly out of the duct or is blocked.* This may be suspected:

   (*a*) because of failure of bile to drain.
   (*b*) because of leakage of bile round the tube.
   (*c*) because of a biliary fistula.

Some cases are due to blockage of the tube by obstructing debris which clears when contrast is injected. But even when the T-tube is genuinely out of place it can be of value to inject, for contrast may still flow into the bile-duct and may show it and the sphincter of Oddi to be normal.

4. *Poor outlining of the duct system.* The concentration of contrast recommended is somewhat dilute so as to avoid 'blotting out' small stones. In some cases the narrowness of the ducts or rapid escape of the contrast lead to poor opacification of the duct system. It is then advisable to tilt the patient head-down and make a forceful injection of 45% Hypaque.

5. *Long cystic duct remnant.* The conventional AP and oblique views do not always show the remnant of the cystic duct because it may overlie the bile-duct. A high oblique or a lateral view may be needed to show this adequately; the use of the lateral view in T-tube cholangiography is discussed by Stern, Schein and Jacobson (1962).

Reference

STERN, W. Z., SCHEIN, C. J. and JACOBSON, H. G , (1962) *Am. J. Roentg.*, **87**, 764.

APPENDIX

Trolley setting

Sterile—upper shelf

1 towel; gauze and wool swabs.
1 Gallipot for Hibitane spirit.
1 × 20 ml. syringe and drawing-up cannula.
Needle—hypodermic size No. 1.
1 pair artery forceps.
Arteriographic flexible connection.

Lower shelf

3 ampoules of 20 ml. 25% Hypaque and 3 ampoules 45% Hypaque (in bowl of hot water).
Bottles—chlorhexidine spirit
        —100 ml. normal saline.

CHAPTER 17

DACRYOCYSTOGRAPHY

(*William Campbell, M.D., Edin., D.M.R.D., London*)

THE weeping eye is a surprisingly common affliction. Under normal circumstances the tears which are secreted by the lacrimal gland are evaporated from the globe at the rate at which they are produced, and any slight excess is drained away through the lacrimal passages. Weeping, or the spilling over of the tears from the conjunctival sac on to the cheeks, is due to one of two causes:—

1. Excessive secretion of tears so that the lacrimal passages cannot drain them away. This is called lacrimation.
2. Inadequacy or blocking of the lacrimal passages so that they cannot accommodate the normal secretion of tears, which then spill on to the cheek. This is called obstructive epiphora.

The weeping of obstructive epiphora is much the commoner condition, and of the methods of investigation of the cause of epiphora, radiology, i.e. dacryocystography, is the most valuable. It determines the site, and sometimes also the nature, of the obstruction to the flow of tears and so helps to decide the type of treatment for the relief of watering.

There are four main causes of obstructive epiphora:—

1. The puncta may be misplaced or abnormal so that tears cannot enter them.
2. The lacrimal passages may be blocked at some point by inflammation, neoplasm, trauma, atresia or foreign body.
3. The nose may be obstructed.
4. The passageways, although permeable, may be functionally inefficient.

Anatomy of the Lacrimal System.—To obtain and to interpret a dacryocystogram a knowledge of the anatomy of the system is required, (Fig. 14). The most important anatomical point to note with regard to the technical procedure is that *the lower canaliculus has two different portions—a vertical portion and a horizontal portion*. The vertical part is approximately 1–2 mm. long, whilst the horizontal part is approximately 7 mm. long.

The canaliculi converge into an ampulla called the sinus of Maier, from which the tears flow through flaps of mucosa, called valves—the valve of Rosenmuller and the valve of Hauske—into the lacrimal sac. The lacrimal sac, which lies on the lateral aspect of the nasal bone, is 1 cm. long, and it terminates at the constriction of the split fascia of the orbicularis muscle which surrounds and indents it. At this level there is also a fold of mucosa called the valve of Krause. The duct continues to widen as it progresses downwards through the bony rim of the canal. In the middle of the canal there is a further constriction in the duct called the valve of Taillefer. The duct empties into the nose below the inferior turbinate bone through the valve of Hasner.

Four further points should be noted with regard to the anatomy.

Firstly, the puncta are not visible until the lids are everted. Secondly, the upper punctum is medial to the lower. Thirdly, the canaliculi enter the antero-lateral aspect of the lacrimal sac close to its apex. Finally,

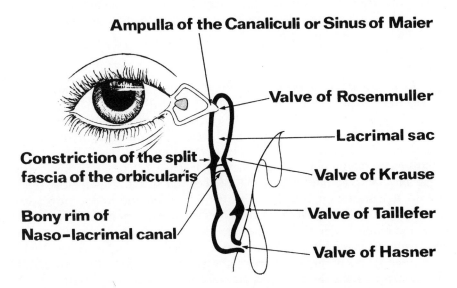

Fig. 14. The radiological anatomy of the normal lacrimal system.

the opening at the lower end is further back in the anatomical specimen than would appear to be the case from the radiographs.

Contra-indications.—Acute infections of the eye and/or periorbital tissues.

Anaesthesia and premedication.—In the child, the examination must be carried out under general anaesthesia. In the adult, the examination is performed on the outpatient with or without local anaesthesia. If a local anaesthetic is used, the patient must be warned of the danger of dust or grit blowing into the eye when leaving the building, or he should be detained for a short time while the anaesthetic is wearing off. Ophthaine is the local anaesthetic of choice. It tends to prevent the reflex watering which is caused when the procedure is attempted without the help of anaesthesia and makes for co-operation by the patient in that he blinks much less. Local anaesthesia is not, however, essential. No premedication is required.

Summary of the procedure

1. Any fluid contained in the lacrimal sac is expressed with a finger pressed against the side of the nose.
2. Local anaesthesia to the conjunctival sac.
3. The lower lacrimal punctum is dilated.
4. The lacrimal cannula, attached to the syringe, is passed along the lower canaliculus to the wall of the nose and then withdrawn slightly.
5. Contrast is injected until it is tasted in the mouth or regurgitated from a canaliculus.
6. Films are taken:—40° occipito-mental and a true lateral of the face, both with the patient lying in the prone position and centring through the infra-orbital margin.

Procedure

1. *Radiographic equipment.*—Adequate films can be obtained with conventional equipment but macro-radiography is the ideal way to examine these structures. The writer's arrangement of apparatus, illustrated in the diagram (Fig. 15), combines the advantages of simplicity and speed of manoeuvre. The system is made up of a micro-focus tube, a light-beam diaphragm and a table which has a perspex top ruled with cross lines, so that the tube can be pre-centred to the lines on the table and to a cassette on the floor below. The table-top-to-floor distance relative to the tube-to-table distance is such that a magnification of $2\frac{1}{2}$ times is obtained. The table is the same height as the radiographic couch and two of its legs have wheels, so that it can be run up against the couch. The latter takes the greater part of the patient's weight. The placing of the cassettes on the floor is facilitated either by fixed marks on the floor

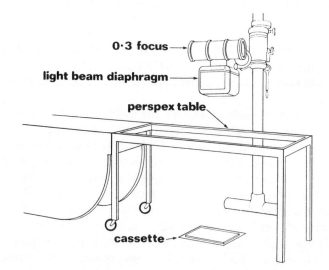

Fig. 15.—Dacryocystography. Apparatus for macro-radiographic technique.

or by using a 12″ × 10″ sheet of yellow paper, Sellotaped into the position indicated by the light beam. The radiographs are taken at 125 kV., at an average time of 1·5 seconds for the P.A. view, and 0·15 seconds for the lateral view, using 12″ × 10″ films and standard screens. The air-gap filtration is sufficient to cut out the scattered radiation as efficiently as a diaphragm and high definition films are produced.

2. *Other equipment.*—A blunt Nettleship dilator is required and a 2 ml. syringe fitted with a silver lacrimal cannula of the Luer-lock type. This type of needle cannot be blown off if excessive pressure is applied, an important point when working above the cornea. The sterile disposable syringes now available give such an excellent grip to the cannula, if firmly applied, that if they are employed it is no longer essential to use a special syringe of the Luer-lock type.

A good light is essential to achieve a successful cannulation.

3. *Contrast medium.*—The opaque medium of choice is Neo-hydriol fluid.

Technique

General Comment.—The sequence of events is to inject the lacrimal system of one eye with the patient supine. He is then turned over and a 40° occipito-mental film obtained followed by a true lateral (still lying in the prone position). The central ray is directed through the infra-orbital margin. The process is repeated for the other eye if indicated.

With the orbito-meatal base line at 40° to the table top, the lacrimal duct is virtually parallel both with the table and the film below, so that it is magnified without distortion. True lateral projections can be used instead of obliques because the high kilovoltage penetrates both ducts and if both eyes are injected one side can be distinguished from the other.

Detailed description.—The patient lies supine with the head resting on a pillow. An important preliminary manoeuvre is to express any fluid which may be distending the lacrimal sac by applying pressure with the finger over the sac against the nose. (This fluid usually appears through the punctum or may go down into the nose. Its removal stops confusion from globulation of the oily medium in the sac. This could be mistaken for a cystic state or multiple diverticula of the sac. The fluid expressed may be purulent or serous.) Next, two or three drops of proxymetacaine (Ophthaine-Squibb) are instilled into the region of the lacrimal lake. Whilst the anaesthetic is taking effect 2 ml. of contrast are drawn into the syringe, the cannula is firmly attached and the air-bubbles are expelled.

The patient is directed to look upwards and outwards and the lower punctum is displayed by depressing the lower lid with the pad of the thumb. With the punctum clearly displayed, a Nettleship dilator is placed vertically into it and then rotated through 90° to follow the bend between the two portions of the canaliculus and to dilate the punctum sufficiently to accept a silver lacrimal cannula. The diagrams illustrate this manoeuvre (Fig. 16). The Nettleship dilator is then quickly exchanged

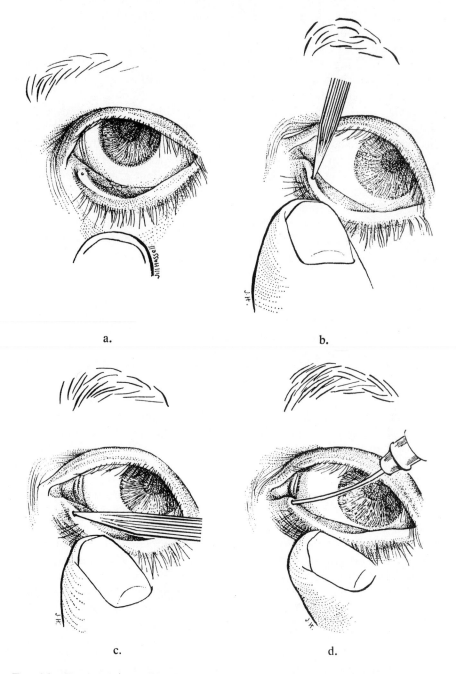

a. b.

c. d.

FIG. 16.—Dacryocystography—technique of cannulating the lacrimal canal. *a.* The lower lid is everted by traction with the thumb. *b.* The lacrimal dilator is inserted vertically into the punctum. *c.* The dilator is then turned through 90° and directed medially. *d.* The lacrimal cannula is inserted in a similar manner and ·passed medially until it strikes the wall of the nose. It is then withdrawn slightly.

for the loaded syringe and cannula. The cannula is then introduced vertically through the punctum into the canaliculus, turned through 90° into the horizontal portion and slid gently along it until the lateral wall of the nose is felt with the cannula. During this procedure the canaliculus is held on the stretch by the pad of the thumb so that the vertical and horizontal portions are drawn into a straight line and the risk of raising and perforating a fold of mucosa is eliminated. When the nasal wall has been felt, the needle is withdrawn slightly and the injection made. As soon as the patient tastes the fluid he can be turned over on to his face and the routine projections taken. It is unnecessary to place droplets of contrast on the eyelashes to mark the position of the eye since the high definition of the films produced by this method shows the soft tissue outline of the lids themselves. Furthermore, droplets on the lashes tend to confuse the shadows of the canaliculi and sac in the lateral projections.

If, when the injection is being carried out, regurgitation takes place through the upper canaliculus, it is then worth while occluding the upper punctum with the Nettleship dilator and injecting the lower canaliculus again to obtain complete filling of the lacrimal duct or what may remain of it. On the other hand, if regurgitation occurs from the lower punctum it is worth while occluding it and injecting the upper canaliculus. The upper canaliculus is not so easily entered as the lower. Under these circumstances the fluid may not be tasted and radiographs should commence when a small amount has regurgitated.

The fluid Neo-hydriol passes through the normal duct in 15–30 seconds. Radiography should therefore be carried out expeditiously after the injection in order to demonstrate as much of the system as possible. This applies particularly to those cases of epiphora due to an incomplete obstruction.

With the macro-radiographic technique the film size 12″ × 10″ ensures that the floor of the nose is seen in both projections. In the lateral view this size makes certain that the naso-pharynx is visualised so that there can be no doubt about the transmission of medium from the duct. The radiographs resulting from the investigation should demonstrate three points:—

1. The size, shape and position of the parts of the lacrimal system.
2. The level of obstruction to the flow of tears.
3. The degree of the obstruction.

Special problems.—If a stricture of the lower canaliculus is encountered with the injecting cannula the injection should be carried out through the upper punctum. If the lower punctum is too small to be entered, then a search should be made for a fistula of the canaliculus and the injection made through this, if it exists. If no filling of a swelling on the medial side of the nose is achieved by means of an injection through the canaliculus, it is probably due to a lacrimal cyst or diverticulum and it should be possible to demonstrate its size by aspiration through the skin and injecting with Neo-hydriol through the aspirating needle. However, the possibility of a mucocele of the ethmoidal or frontal sinuses, or of a meningocele, must not be overlooked.

Hazards and complications.—The main hazard of the procedure is that of creating a false passage by faulty technique when inserting the cannula. Unless the two directions of the canaliculus are appreciated and the cannula rotated through a right angle whilst inserting it, at the same time holding the canaliculus on the stretch, it may be easily ruptured. If this happens the periorbital tissues are injected with the opaque medium; it takes approximately three months for the Neo-hydriol to be absorbed.

A further danger is that of damaging the punctum with the dilator. The operator must be sure that the dilator is *smooth*, and only reasonable pressure should be applied during dilation.

Another hazard can be avoided by remembering the delicate nature of the cornea over which the instruments may pass, and taking care not to touch it.

Finally, the labels on the solutions, particularly the solution of local anaesthetic, should be read with care before allowing them to enter the eye.

APPENDIX

Trolley setting

Sterile—upper shelf

Nettleship dilator (blunt).
Silver lacrimal cannula—Luer-lock hub.
2 ml. Luer-lock syringe or sterile disposable syringe.
Swabs. Sterile towel.
Cannula for drawing up contrast.
Gallipot and kidney dish.

Lower shelf

Neo-hydriol fluid (ampoule in warm water).
Bottle of Ophthaine (proxymetacaine hydrochloride—Squibb).

Also required

A bright light may be needed if the room lighting is inadequate.

Editor's note. Readers interested in the diagnostic aspects of this examination should consult the author's paper:
CAMPBELL, W. (1964) *Br. J. Radiol.*, **37**, 1.

CHAPTER 18

SIALOGRAPHY

THIS is a long established procedure but Rubin and Holt (1957) have added the 'emptying film', taken after stimulation of the gland, so that it may provide a demonstration both of the structure and the function of the gland.

Contra-indications.—Oral, parotid or submandibular infection.

Preparation.—Any false teeth are removed.

Preliminary films.—These confirm satisfactory centring and exposure and may reveal opaque stones, especially in the submandibular region.

Opaque parotid stones are less common and less densely opaque and they show best on a 'soft' AP film; this may be obtained by placing a 'non-screen' film in front of the cassette used for the plain AP film. If the plain lateral film before parotid sialography shows many opaque fillings in the upper teeth, the lateral film following injection may be taken with the tube angled down about 15° to throw the fillings clear of the duct in addition to an oblique angled upward. The standard projections are:—

A.P. and lateral for the parotid region (lateral-oblique for the second side when both glands are examined).

A.P. and lateral-oblique for the submandibular region.

The examination is usually performed with the patient lying on a horizontal couch. When carried out sitting it is helpful if an assistant holds the head.

Contrast media.—The medium of choice is Lipiodol Ultra-fluid. Myodil is equally opaque but drains more rapidly and should be used with a 'closed'[1] system. Water-soluble media are less opaque and we have not obtained such good results with them as with Lipiodol. Hypaque 85% or Conray 420 are the most satisfactory watery media. They too should be used with a 'closed'[1] system.

Summary of the procedure

1. Plain films—A.P., soft-tissue A.P., and lateral (oblique for the submandibular gland).
2. Find and dilate duct orifice.
3. Introduce catheter and inject contrast until pain is experienced.
4. Inject up to 0·2 ml. more, then close tap and take films.
5. Give lemon to suck, take lateral (oblique) 5 minutes later.

---

[1] With a 'closed' system the cannula or catheter used for introducing the contrast is kept in the duct while the films are being taken and contrast is not allowed to escape.

Equipment

1. The duct may be cannulated or catheterised. Lacrimal cannulae are satisfactory for small orifices; the larger tapering sialography cannulae are suitable in the majority of cases. The method recommended here, however, employs a finely tapering polythene catheter and the use of cannulae is only suggested when a more rigid system seems necessary, e.g. in order to enter a very mobile or irregular orifice. The catheter employed is a polythene PE or OPP 160 drawn out to a fine tapering point, Fig. 17, and fitted with a tap/adaptor (see Appendix for details of preparation).

2. An Anglepoise lamp or other adjustable source of bright light is essential.

FIG. 17.—Polythene catheter used for sialography.

Procedure

1. Although absolute sterility is impossible the operator scrubs up to ensure that everything put into the mouth is sterile. A sterile towel is laid over the patient's chest and the adjacent table. A 5 ml. syringe is filled with contrast and attached to the catheter. Air bubbles are expelled and contrast driven to the tip of the catheter; the tap is then closed.

2. **Catheterising the duct.**—*Parotid.*—The orifice is inside the upper lip, level with the second molar in a small fold of mucosa. It may be clearly visible; if not the inside of the cheek is dried with a swab, the parotid gland is massaged and the region of the duct orifice is watched for the efflux of saliva. With diseased glands it may be necessary to give a stimulant such as a lemon slice, dilute vinegar or $\frac{1}{2}\%$ HCl to be sucked for a few moments. When the orifice is localised a Nettleship's lacrimal dilator is slid *gently* into it and rotated.

After this a series of lacrimal probes are passed into the duct as far as the masseter. (Sometimes one may feel a stenosis--this should be noted in the report.) Dilatation is continued only until the catheter will fit snugly. When the duct orifice has been found and dilated the catheter is carefully introduced and advanced until the taper is *firmly* gripped by the duct. Holding the catheter steady, the operator then instructs the patient to close his lips firmly onto the catheter. The catheter should be disposed so that it does not overlie the area under examination in either view—taping it to the patient's nose is a good way of doing this.

Submandibular.—The principles of catheterisation are the same but it is usually more difficult to find the duct orifice. It lies in the fleshy sublingual papilla just lateral to the frenulum. The patient is asked to put his tongue against the roof of his mouth and then to open the mouth widely. This position is difficult to maintain without moving and it may be necessary for the patient to insert a finger to steady the tongue. Occasionally his other hand may be used to push under his chin to lift the floor of his mouth. The orifice is sought in a good light, drying the papilla with gauze so that the salivary efflux shows clearly. A pair of non-toothed dissecting forceps may help to steady the papilla. Probing is continued gently and patiently until the lacrimal dilator slips into the duct mouth. The essence of the manoeuvre is to spend time on showing where the orifice is, rather than on probing blindly. Gentleness is most important.

Injecting the contrast

Before contrast is injected the X-ray tube is centred for the A.P. film. The patient is told that he will feel a pain or discomfort in the face and asked to indicate this by lifting a hand. The syringe is gently squeezed until this signal is seen; about 0·2 ml. further is injected, the tap is closed and the syringe laid down. The film is then exposed. Now the patient turns gradually into position for the lateral (or oblique) film. The tube is again centred, a further 0·1–0·2 ml. of contrast is injected and the film exposed as before. The catheter is disengaged and replaced on the sterile trolley. If the patient feels no pain, up to 2 ml. may be injected but a film should then be taken to check filling.

When processed, the films are inspected. If they are technically satisfactory the patient is given a half-lemon (or an 'acid-drop') to suck and further A.P. and lateral films are taken 5 minutes later to evaluate emptying of the gland.

Extra films.—When a tumour is present additional information about displacement or deformity of the gland is gained from a submento-vertical view or occipito-mental with 30° added tube tilt towards the feet. There is no need to maintain a closed duct system for these views, unless water-soluble media are employed.

Difficulties

Catheterising the duct.—Provided the duct orifice is adequately dilated and the catheter tip is not too long or too flexible there is seldom any problem in inserting the catheter into the parotid duct. Sometimes it helps to straighten the duct by drawing the cheek forward. If it proves impossible to use a catheter, a suitable, e.g. lacrimal, cannula attached to a flexible connection should be tried. This is held in position during the injection with a finger placed over the duct orifice; the finger is removed just before the film is exposed. Difficulty is particularly likely to occur when attempting to catheterise the sub-mandibular duct. The patient is asked to lift the tip of the tongue against the roof of the mouth and the duct region is swabbed repeatedly so as to achieve complete dryness. Massaging the submandibular region will then produce some efflux of saliva. It has been found that a Nettleship dilator with the tip bent through a right-angle is particularly useful for entering the duct.

Displacement of the catheter from the duct orifice.—This is most likely to occur with a wide duct orifice and can be suspected when the patient fails to give any sign of pain after $\frac{3}{4}$–1 ml. of contrast have been injected. If inspection of the mouth shows that the catheter is no longer in position it is re-inserted as firmly as possible; it may need to be held in position with a finger during injection. It is therefore as well to keep a few wider catheters, i.e. PE 205 or PE 240 similarly prepared for injection of patients who have unusually wide ducts.

Underfilling of the gland.—In the absence of organic narrowing underfilling of the gland is usually due to failure of the catheter to remain in the duct, as noted above. Grossly sialectatic glands may take larger volumes without pain but it is never wise to inject more than 2 ml. without taking a film to show the degree of filling which has been produced.

Obscuring of the duct by teeth.—Sometimes the parotid duct is masked by teeth, particularly by opaque fillings. Oblique films angled upwards or downwards may then be needed to show the duct clear of teeth.

Complications.

1. **Overfilling of the gland.**—This occurs if injection is continued excessively after the patient begins to feel pain. It is unavoidable in a few cases because some patients are relatively insensitive to distension of the gland. It is undesirable because the patient suffers and because the fine acinar structure of the gland is obliterated. Sometimes the gland becomes swollen and painful for some hours; symptomatic treatment with warm or cold compresses, gentle gland massage and analgesics may be needed.

2. **Rupture of the duct.**—This is a rare complication using the catheter technique. It results from trauma during duct dilatation and causes little if any pain. If it is known to have occurred the examination is abandoned for two weeks. If it is suspected but not definite, injection of a water-soluble medium may be tried. If the duct is not intact injection will rapidly cause pain.

3. **Introduction of infection.**—Unless sialography is performed in the presence of oral infection or with a considerable amount of trauma infection of the gland is very unlikely.

References and further reading

ROSE, S. S. (1950). *Postgrad. med. J.*, **26**, 521.
RUBIN, P., and HOLT, J. F. (1957). *Am. J. Roentg.*, **77**, 575.

APPENDIX

Trolley settings

Sterile—upper shelf

1 towel; gauze and wool swabs.
1 × 5 ml. syringe with cannula for drawing up contrast.
1 pair tissue forceps.
2 lacrimal duct dilators.
Assorted lacrimal duct probes.   } in kidney dish with Hibitane/spirit.
Lacrimal needles; sialography needles.
P.E. 160 Sialogram catheter (see below).
P.E. 160 Tap/adaptor.

Lower shelf

Lipiodol Ultra-fluid—10 ml. ampoule in warm water—or ampoule 85 % Hypaque.
Slices of lemon.
Adhesive plaster strips for catheters.
Sterile packs—PE 205 sialogram catheter.
                   „   „   tap/adaptor.
Novex 3-way adaptor.

Also needed

Anglepoise or other bright light.

Preparation of catheters

For most ducts polythene tubing size PE or OPP 160 is a suitable diameter. To make a suitably tapered tip for the catheter it is held over a *low* flame and drawn out when it has become softened. A spirit lamp is commonly employed with the flame as small as possible. The tubing is held between the two hands a short way above the flame and is rotated continuously until it becomes softened. It is then removed from the region of the flame and pulled apart gently. If the softening has been correctly judged it will draw out to a suitable taper with a long waist in the middle. The tubing is allowed to cool and the tapered region is cut across at a suitable point and inspected. If it is satisfactory a length of about 30 cm. of tubing is cut and the other end is flanged, as described on page 239. This technique demands some practice but if care is taken it can usually be mastered. Once the 'knack' has been acquired it is advisable to make a large batch of catheters. As already noted larger catheters made with PE 205 tubing should also be prepared though only a few of these are needed.

CHAPTER 19

BRONCHOGRAPHY

THERE are a number of methods for the introduction of contrast media into the bronchi. Any of them can give satisfactory results when their technique has been mastered and each has its especial advantages. We strongly recommend the radiologist to master two entirely different techniques so that the situation can never arise where the examination cannot be carried out because no operator is available skilled in the only method suitable for a particular case. We suggest the trans-nasal and the crico-thyroid routes since both are satisfactory methods which differ widely and hence are the most complementary.

The trans-nasal route is very widely practised; it is safe, it requires little special skill and it must be regarded as the method of choice for the un-supervised beginner. The crico-thyroid route, *in experienced hands*, is quick, certain and often preferred by patients who have experience of both methods. However, it carries the possibility of certain complications which cannot result from the trans-nasal approach and in general it should be reserved for more experienced operators, for use in special circum-stances or when the trans-nasal route has failed.

Before discussing the special points of the various methods the features which they have in common will be considered.

Contra-indications

1. Acute chest infection or upper respiratory tract infection.
2. Heart disease of more than mild degree.
3. Emphysema and low vital capacity; asthma. When chronic bronchitic or asthmatic patients require bilateral bronchography it is inadvisable to examine both sides at a single investigation.
4. Any recently consumed food or fluid may be vomited so that a meal or drink taken by mistake is an absolute contra-indication.

Preparation

1. When bronchiectasis is clinically likely or when the patient is producing copious sputum, postural drainage and percussion are carried out for at least 3 days before examination and preferably for a week if bronchiectasis is at all severe. A simultaneous course of antibiotics results in a marked—temporary—improvement in the amount and quality of the sputum.
2. The patient must not eat or drink for at least 4 hours before the examination.

Preliminary medication.—Bronchography is a successful and 'smooth' procedure in some patients without any premedication, but it is not an agreeable experience and some sedation is desirable. The ideal preparation would blunt the patient's apprehensions while leaving him co-operative; it would suppress coughing and it would dry up secretions—bronchial, buccal and naso-pharyngeal. The latter is most important, for the surface anaesthetics used are much less effective if the mucosae are 'streaming', and atropine 0·6 mg. or scopolamine 0·4 mg. intramuscularly are given three-quarters of an hour before the start of the examination. In-patients are also given pethidine 1·5 mg./kg. of body weight. During the period while awaiting the examination oral anaesthesia is started (see below) except for those patients who are to be examined by the crico-thyroid route. All patients are given Valium (diazepam) 10–15 mg. intravenously at the outset of the examination.

Preliminary films.—The quality of film needed is darker than the ordinary P.A. chest film. If a grid[1] is used the kilovoltage is about 10–15 kV. higher than for a normal chest X-ray. It should be possible to see the translucency of the trachea clearly on the P.A. film but the lung fields must not be blacked out. The lateral film needs to be only slightly darker than an ordinary chest lateral; here a grid must be used.

When the right side only is to be examined a P.A. and right lateral film are always taken; where examination is of the left side alone, a P.A. and left posterior-oblique. A left lateral may also be taken although only the first two positions are essential. Similarly a right posterior-oblique (RPO) is a valuable third film for the right side and often helps to clarify problems of bronchial anatomy. If the bronchogram is to be bilateral always inject the right side first so that a right lateral may be obtained without its being obscured by the filled left bronchial tree. A lateral of the left side is not necessary and contrast medium in the right lung will not obscure the left side on the left posterior-oblique film.

Fast films and screens are used to reduce exposure times.

Choice of local anaesthetic.—Lignocaine 4% for topical use has been employed for many years for anaesthesia of the larynx and trachea. The toxicity of this type of local anaesthetic is greatly reduced when lower concentrations are used and for intratracheal instillation a 2% or even 1% solution is preferable. A maximal amount of 5–6 ml. of 4% Lignocaine is generally recommended for an average adult patient but much of this can, of course, be given in a more dilute form. An alternative topical anaesthetic with a higher maximal dose is Prilocaine (Citanest–Astra Hewlett). Up to 10 ml. of 4% Prilocaine may be used for tracheo-bronchial anaesthesia though it should also be diluted to 2% for use within the trachea and main bronchi. Its advantage is that it is possible to give further anaesthetic if the initial anaesthesia appears unsatisfactory.

Contrast media.—The generally accepted medium of choice at the

---

[1] The use of a grid is a matter of individual choice and the radiologist will discover for himself whether or not he prefers grid films.

moment is propyliodone (Dionosil-Glaxo). The oily variety is more often used because it is less irritating but the arachis oil in which it is suspended does not always disappear and may cause granulomata. There is however no evidence of actual harm resulting therefrom.

An alternative medium is Hytrast—an aqueous suspension of Iopydol and Iopydone. This contrast medium has a higher iodine content (50%) than Dionosil and is therefore more densely opaque. It coats the bronchi more tenaciously and even after coughing adequate bronchial outlining may persist. It is more irritating than Dionosil and more prone to cause coughing when first instilled. Because of reports of damage to the lung and particularly the formation of granulomata an earlier form of this medium was withdrawn from use in the United States. However one of us, (H.M.S.), has used the current medium regularly for the past four years without encountering any clinical difficulties but as with all relatively new contrast media it is important to be alert for evidence of harm resulting from the medium, both in one's own patients and in reports in the literature.[1] One point of great importance in its use is that it is crucial to achieve the correct viscosity of medium. Too dilute a suspension will allow excessive contrast to reach the lung acini, so increasing the likelihood of lung granulomata. Undiluted contrast may be so viscous that if it is used for an examination under general anaesthesia, the bronchi may be blocked by the thick contrast medium, and the lung may fail to expand. The addition of $\frac{3}{4}$–1 ml. of water, carefully measured, to the 20 ml. ampoule, *with very thorough shaking*, will give a mixture of suitable viscosity.

Dionosil may be thickened by adding Dionosil powder—1 or 2 gm. to the 20 ml. ampoule. Although this makes 'atrial filling' less common it is not recommended because of the following disadvantages:—

1. Unless mixing is really thorough—which means vigorous shaking of the bottle for 5–10 minutes—plugs of contrast may form.
2. The thickened contrast may not move out in narrow bronchi and the peripheral lesions of chronic bronchitis will not be shown. In chronic bronchitis moreover, it appears to cause more breathlessness than does unthickened contrast.
3. If coughed up it does not leave such a good coating of the bronchi as with unthickened Dionosil.

For chronic bronchitics and other breathless patients, therefore, un-thickened Dionosil should always be used. Too much, indeed, has been made of the drawbacks of peripheral filling; underfilling distally is also a mistake and the best results are achieved with a well-judged amount of unthickened contrast kept in a bowl of iced water and shaken intermittently until just before use.

The volume needed varies with the size of the patient and the likelihood of bronchiectasis. A large patient will require about 18–20 ml., an average

---

[1] At the time of going to press this medium has been withdrawn by the manufacturers.

patient 14–16 ml. and a small patient 12 ml. to a side. When there are changes of bronchiectasis on the plain film an extra 2–4 ml. a side is added. When the crico-thyroid route is used an extra 2 ml. is added to allow for some coating of the trachea. If a lobectomy has been performed or in a case of suspected lung hypoplasia proportionately less contrast medium is employed. Coughing during the examination is also an indication for further contrast.

When the examination is completed the patient is told to cough vigorously with suitable posturing to clear the lungs of contrast. It is recommended that a physiotherapist should help with this for chronic bronchitics, bronchiectatic subjects and feeble patients. After crico-thyroid bronchography the puncture site should be pressed on firmly during the coughing—otherwise air may be forced into the tissues or a haematoma may develop.

Posture and Injection

It is not normally necessary to alter the posture of the patient during the injection of contrast medium. In most cases the simplest method is to complete the injection with the patient in one position and then use image intensifier fluoroscopy to check adequate bronchial filling before taking the films.

The patient lies on the side to be examined with the feet tilted slightly downwards and the chest supported on the elbow—the so-called 'Roman feast' position (Pridie, 1966) Fig. 18. The advantage of this position is that the upper part of the thorax is not compressed by the weight of the chest and can therefore expand normally. All the contrast is injected in

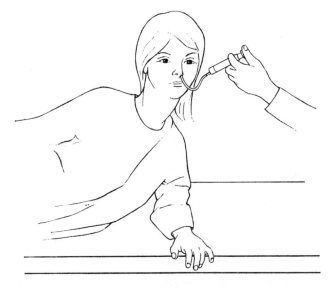

FIG. 18.

this position. When the injection is completed the patient is rolled half way forward (semi-prone) and half way onto her back, taking a deep breath in each position.[1] Filling is then quickly checked fluoroscopically and any necessary films are taken with the under-couch tube. After this the patient gets off the table and films are taken in the *erect* position. If filling of a particular area is seen to be inadequate then the patient should be positioned approximately as outlined below, an extra 3–4 ml. of contrast medium being injected as required to outline the underfilled area. In each such position the patient is supported by an assistant and the position is maintained for at least ten seconds followed by a deep breath. The individual positions are shown in Figure 18. Thus:—

To fill the lower and middle lobes the patient (leaning towards the side under investigation) bends well backward (position 1), then sideways (position 2) and then forward (position 3).

To fill the upper lobe (position 4) she lies on her side with the head elevated and the feet slightly lowered. When injection is completed and after a short pause with the crico-thyroid approach (to allow the contrast to run down the trachea) she is turned half-way onto her face (position 5) and then half-way onto her back (position 6) while the head of the table

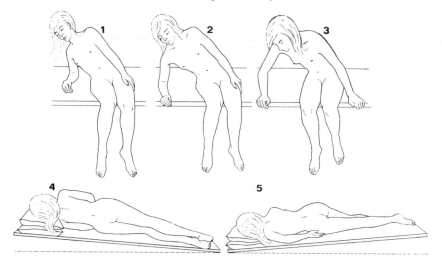

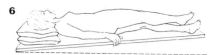

Fig. 19.—Positions for bronchography. 1, 2 and 3 are shown for the right side, 4, 5 and 6 for the left.

---

[1] A further variation on this method has recently been described by Amplatz and Haut, (1970). They inject all the contrast in one bolus while the patient holds her breath in full expiration, lying on her side. The ensuing inspiration draws the contrast out into all the main bronchi. Our limited experience of this refinement has shown that it works well.

is lowered. In each of these positions a deeper slow breath is taken to draw the contrast outward. When it is particularly important to show the upper lobes an extra 2 ml. is injected in position 4.

By this method the bronchi will normally all be filled, but when there is basal bronchiectasis a few more moments of posturing in the position appropriate for the affected lobe may help to fill the dilated—often secretion-filled—bronchi. Prolonged posturing may also be needed to fill a collapsed lobe. It may then be necessary to take further films after 10–15 minutes have elapsed in the hope that the contrast will have moved into the diseased area.

Coughing is sometimes a problem during instillation of contrast, and may be provoked by deep, sudden breaths and rapid alterations of position. Movements should therefore be deliberate and breathing slow and shallow. Pinching the nose, supporting the chest wall and an encouraging flow of talk will help to suppress coughing. Of these the most important is the 'flow of talk'; the practised bronchographer does not stop talking to the patient for more than a few moments—a state bordering on hypnosis is often produced.

The films.—If image intensification is available filling should be checked as described earlier, by screen control. Films are then taken as soon as the area under especial examination has been filled—this ensures that films with adequate contrast delineation have been obtained and will prevent a repetition of the investigation should the erect films prove inadequate. It is, for example, often possible to take a series of spot films of the orifice of a particular bronchus in varying degrees of obliquity and so show that it is either normal or obstructed. The films at fluoroscopy are taken as quickly as possible; when they have been exposed, without waiting for them to be processed, films at 6 feet are taken as well; we prefer to take them erect at a chest stand with a grid. (When they are taken on a table the right lateral or left posterior-oblique is taken first.) The patient is watched carefully throughout and *if there is any suspicion that she has breathed the film is repeated.* When the films have been exposed the operator may prefer to wait to see them; if so the patient lies on the side which has been 'filled'. Otherwise in bilateral examinations the left side may be commenced when the right is completed. It is usually preferable to wait to see the films unless the patient is very restless. When one has ascertained that the films of the right side are satisfactory and that no more contrast or further films are required, the patient may be told to cough up the contrast from the right side. A further 2 ml. of lignocaine 1% are then instilled into the left side.

As already mentioned the standard films are P.A. and right lateral for the right lung, P.A. and left posterior-oblique for the left. An A.P. lordotic view can be used to separate out the basal bronchi in the antero-posterior projection. If the right side has to be re-examined after the left has been filled a right posterior-oblique is taken. After lobectomy or in cases of hypoplastic lung it is advisable to take three views—P.A., lateral and right or left posterior-oblique, since there may be distortion of the normal anatomy. This applies in any other condition where the anatomy may be disturbed, e.g. scoliosis, lobar collapse or fibrosis.

We would particularly stress the importance of carefully checking the films when they have been processed. Each bronchus must be identified so that underfilling of any area will be noted and more films taken after further injection and posturing. Because of the patient's discomfort one may hurry to conclude the examination and find afterwards that it is incomplete. An incomplete bronchogram is seldom of value to the surgeon.

After-care.—1. Postural coughing to bring up the contrast has been mentioned.

2. The patient should be told not to eat or drink for at least 3 hours because of the danger of aspiration through the anaesthetised larynx.

3. She is warned that there may be troublesome coughing when the local anaesthetic wears off, that her sputum may be blood-streaked and that she will have some degree of sore-throat.

4. The operator should assure himself that out-patients have recovered from their premedication and from the effects of local anaesthesia. In most cases it is unwise to allow a patient to drive a car until some hours have elapsed.

THE TRANS-NASAL ROUTE

Summary of the procedure.

Preliminaries: A. The patient is premedicated.
      B. The patient sucks a Xylocaine lozenge.
      C. Preliminary films are exposed.

Procedure: 1. Spray through the nose with 4% lignocaine or prilocaine during slow inspiration.
    2. Apply 'Xylocaine viscous' to the tubing.
    3. Pass the tube through the nose to the naso-pharynx and ask the patient to say a high-pitched 'Eee'.
    4. Inject 3 ml. 1% lignocaine or 2% prilocaine onto the vocal cords through the tube, during inspiration.
    5. Advance the tube in the midline until it passes between the cords; confirm position by asking patient to repeat 'Eee'. Inject 4 ml. 1% lignocaine into the trachea.
    6. Attach tube to cheek. Inject contrast with patient lying on the side under examination. Check filling fluoroscopically.
    7. Expose films. Repeat procedure for the other side if required.
    8. When the examination is complete remove tube and encourage patient to cough.
    WARN THE PATIENT NOT TO EAT OR DRINK FOR 3 HOURS AFTER EXAMINATION.

Nasal route—details of the procedure

1. The intubation is performed with a curved catheter. Size 7 or 8 (English gauge) is best as thin tubing is difficult to control. The tube should be well curved (Fig. 20); too straight a tube may pass down behind the larynx. A simple Jaques catheter or a shortened Ryles tube are employed with a wire within the lumen suitably bent to impart the necessary curvature during intubation (see Fig. 20). A Luer-Record adaptor is sometimes needed to allow attachment of a syringe.

2. **Topical anaesthesia.**—This is initiated by sucking a Xylocaine lozenge or by gargling with 'Xylocaine viscous'. The lozenge is given 25 minutes before the examination starts. The gargle—2 fluid drachms (8 ml.)—is given 15 minutes before the examination and again at the outset. The gargle should not be swallowed. The Xylocaine lozenge is less effective than an amethocaine lozenge but the latter is potentially dangerous and cannot therefore be recommended.

Once the examination commences the patient, wearing a gown and holding gauze swabs, sits facing the operator. A bowl is on a chair beside him. Spraying is carried out through whichever nostril is clearer. The patient is told to inspire slowly and deeply, keeping the mouth closed and compressing the other nostril with a finger. Spraying through the chosen nostril is continued until full inspiration; the patient is then told to hold the breath for 10 seconds and then exhale. The head is tilted backwards to direct all the anaesthetic into the nasopharynx. The cycle is repeated until about 2–2½ ml. 4% Lignocaine or Prilocaine has been used. The advantages of using the nose alone are (i) it is less unpleasant for the patient (ii) anaesthetic is not wasted on the mouth. The pause after inspiration and before exhalation decreases the amount of anaesthetic vapour lost with expiration.

As soon as spraying is completed the patient is warned that once the tube has been put in, he must not cough.

3. The tube tip is smeared with Xylocaine 'Viscous' and passed into the nostril, insinuating it gently to the back of the nose; some manipulation may be needed and occasionally the opposite nostril may be found the clearer one. There are two sources of difficulty in passing a tube through the nose. The first is that the turbinates may be swollen or the septum

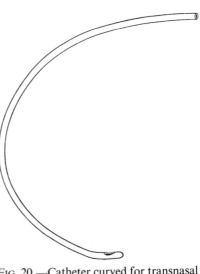

FIG. 20.—Catheter curved for transnasal bronchography.

deviated so that the passage through the nose itself is too small for the tube. In these circumstances it may be necessary to anaesthetise the opposite nostril. When the tube reaches the back of the nose and impinges on the posterior wall of the nasopharynx it is not always smoothly deflected downwards and may be held up at this point. If gentle probing fails to produce onward passage of the tube it should be withdrawn and its tip curved slightly (see Fig. 20). By this means it is usually possible to persuade the tubing to pass into the nasopharynx. When the tube has been passed into the nasopharynx the patient is asked to say a *high-pitched* 'Eee'; this serves as a control with which to compare phonation later

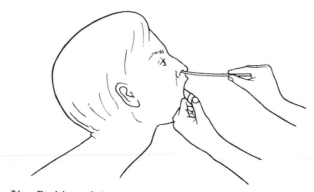

FIG. 21.—Position of the patient during intubation of the larynx. Note that the tongue is pulled well forward.

when the tube is in position between the cords. Now the patient takes up the position shown in Fig. 21, leaning forward with the chin thrust out and the mouth open. The tongue is grasped with a swab in the left hand and pulled forward firmly. A light—torch or headmirror—is shone into the mouth; if a torch is used an assistant directs the beam over the operator's right ear.

The tip of the tube becomes visible just below the palate. When it does a syringe with 3 ml. of 1 % lignocaine is attached to the other end and the patient is asked to breathe slowly and deeply. The lignocaine is injected steadily during inspiration—this injection should drop onto and through the vocal cords and may cause some coughing. The tube is now advanced. It should then be seen to go down in the midline. The patient continues to breathe slowly and deeply. If all goes correctly the tube slips into the trachea after one or two gentle advances. This may cause a burst of coughing with puffs issuing from the outer end of the tube which serve to confirm its position in the trachea. Further to confirm this—for it may also go down the oesophagus—the patient is asked to say 'Eee' again. When the tube lies between the vocal cords the patient may not be able to say 'Eee' at all and if he can the sound is strikingly altered. We have found this a very reliable sign. The guide-wire is then withdrawn. Another manoeuvre to confirm the tube position is to ask the patient to take big breaths in and out with the mouth open. This produces a hissing rattle in the tube and a wisp of cotton-wool held near its end will move with the air currents; alternatively if the end is dipped into fluid during *expiration,* bubbles will emerge. The mouth should not be closed for this as the pressure in the tube can then rise even when the tube is in the oesophagus.

When the tube does not pass easily into the trachea the following points should be considered:—

(*a*) Check that the tube is advancing in the midline, if necessary twisting the tube to bring it centrally. Ask the patient on which side he feels the tube, if at all.

(*b*) Make sure that the patient has not recoiled from the chin-forward position, and that the tongue is pulled well forward.

(*c*) Sometimes the tube will pass through during coughing or, to be exact, in post-tussive inspiration.

(*d*) Try gently rotating the chin to each side, advancing the tube anew with each change in position.

(*e*) The tube used may be too supple or have lost its curve. When this is thought to be the case it is usually possible to restore the curve by inserting a Seldinger guide-wire and bending the assembled guide and tubing so as to restore the curve. Often a small further deflection at the tip of the tube will be helpful in directing it anteriorly into the larynx.

(*f*) When there is no success through one nostril the opposite one should be tried as it may direct the tube more suitably.

(*g*) Anaesthetic textbooks suggest manual displacement of the larynx to one side or the other. We have not tried this ourselves but it seems worth mentioning.

Once the tube is in position it is strapped to the nostril; then 3–4 ml. of 1 % lignocaine is injected into the tube and pushed through by the injection of air. Contrast may then be injected in the posture already described. A small injection of air is needed at the end of injection to clear the tube which is then spigoted.

CRICO-THYROID ROUTE

Summary of the procedure

Preliminaries: A. The patient is premedicated.
　　　　　　　B. Plain films are exposed.

Procedure: 1. With the patient supine and head extended, local anaesthetic is injected into the skin over the crico-thyroid membrane and into the membrane itself.

2. An intramuscular needle on a 2 ml. syringe is passed through the crico-thyroid membrane and 4 % lignocaine

injected rapidly into the trachea. Needle withdrawn immediately.

3. With patient sitting a further 3 ml. 1 % lignocaine is slowly
injected; patient leans towards side under examination. With
patient supine the crico-thyroid needle or needle-catheter
is passed into the trachea and contrast injected in the usual
position.

5. Needle is withdrawn, posturing is completed and the films
are exposed.

NEVER INJECT INTO THE TRACHEA WITHOUT WITHDRAWING
AIR TO CONFIRM THE POSITION OF THE NEEDLE OR CATHETER.

Crico-thyroid route—details of the procedure

The special points about this route are:—

1. The syringe specifically designed for crico-thyroid bronchography is
made for one-handed injection and its short needle locks onto it. When
it is filled with contrast a space must be left between the plunger and the
top of the barrel so that there is room to withdraw the plunger when
confirming that the needle is in the trachea. Other types of syringe may
also be used for this purpose but it is then more difficult to draw back air
into the syringe with one hand so as to check that the needle is correctly
sited. This does not matter however if a needle-catheter (e.g. Longdwel 6719)
is used for the puncture. Once aspiration of air through the central
needle confirms satisfactory puncture, the needle is angled towards the
feet and the catheter advanced over it and well into the trachea.

2. **Topical Anaesthesia.**—The patient lies with pads under his neck to
produce hyperextension. The neck is painted locally with iodine and
then more widely with Hibitane/spirit. The operator runs his index finger
down the central ridge of the thyroid cartilage until it meets the transverse
bar of the cricoid, the gap between the latter and the inferior border of the
thyroid which now lies beneath the finger is the crico-thyroid membrane.
This gap is normally quite easily felt but sometimes a little care is needed
to define it.

A wheal of local anaesthetic is raised over the site of the membrane and
infiltration is then carried deeper. A wide intramuscular needle is now
attached *firmly* to a 2 ml. syringe with 2 ml. of 4% lignocaine or prilocaine
and this is advanced steadily through the membrane into the trachea with
the tip inclined slightly towards the patient's feet. The plunger is drawn
back to confirm—by the bubbling back of air—that the needle is in the
trachea. Then the patient is told—'In a moment you will want to cough—
on no account do so until I tell you'. The needle is grasped close to the
skin so that (i) it can be pulled out rapidly (ii) to enable the depth of the
laryngeal airway to be gauged. The lignocaine is injected rapidly and the
needle is pulled out immediately in a single continuous movement. The
patient is told that he may sit up and cough. When the coughing has

subsided and with the patient sitting, 3–4 ml. of 1 % lignocaine is injected after repuncturing the crico-thyroid membrane. Again the position of the needle is checked by drawing back air and the lignocaine is slowly injected with the patient leaning first to one side and then—if both sides are to be done—to the other.

The skin at the puncture site is nicked with a small scalpel blade— using the blade by itself is less alarming—so that the needle can be easily inserted and will not block with a plug of skin.

3. A useful addition to the needle is a small adjustable guard to prevent its going in too deeply.

4. With the neck still hyperextended on pads, the thyroid cartilage is steadied between the thumb and forefinger of the left hand. The syringe of Dionosil is held like a pencil in the right hand and with the wrist resting on the chest is thrust slowly down through the crico-thyroid membrane. This should be a controlled thrust so that one does not suddenly jerk the needle too far in. Puncture of the membrane yields a definite, slightly grating 'give'.

5. When the needle is thought to have penetrated adequately the left hand takes the barrel of the syringe while the right draws back on the plunger—free bubbling back of air confirms that the needle is in the trachea. *This manoeuvre must be repeated before injection* and after each shift in the patient's position or if injection meets with any resistance. (One should try to draw back exactly 1 ml. at a time so that when one subsequently injects, the amount injected and the amount left in the syringe are known.) When a needle guard is used it is now locked on to the needle at skin level.

6. When a needle-catheter is used the assembled needle and catheter on a syringe are advanced through to the approximate depth of the trachea, if necessary with a rotary movement. Once the aspiration of air has confirmed successful puncture, the needle-catheter is angled towards the feet. Then the butt of the needle is held fixed while the catheter is slid off the needle and down the trachea as far as it will go. It is usually possible to aspirate air to check that the catheter is satisfactorily placed but if its tip is pressed into the tracheal mucosa gentle probing with a P.E. 160 guide wire may be needed to prove that the tip is free in the lumen; *this is essential* before injecting further local anaesthetic or contrast. Once these preliminaries are completed, the catheter is firmly strapped to the skin.

7. The patient is now rolled onto the side to be examined, i.e. onto his right side for the right lung. The head must be held in the extended position during turning of the body lest the needle be displaced. The head is supported on pillows or pads and is angled upwards so as to prevent spilling of contrast upwards through the larynx. As described earlier, the foot of the table is lowered slightly. The needle position is checked again by the withdrawal of air and with the needle point inclining caudally all the contrast medium is then injected. Bronchial filling is checked by image-

intensifier fluoroscopy and if complete the needle is withdrawn. If filling is inadequate the following procedure is undertaken:—

While the operator holds the syringe and needle to prevent any accidental shifts, the patient sits up gradually, swinging his legs over the side of the couch. He is supported when leaning backward by an assistant and when leaning forward by putting his arm around the operator's shoulders. The operator continues to steady the needle with his left hand. A further 2–3 ml. is injected with the right hand in any of the positions 1, 2 or 3 which may require extra filling of a particular segment. Rather longer is allowed for each position than with the tube methods to allow the contrast to run down the trachea. For position 4 the patient lies on the side under examination and the head is elevated on a pillow. At this stage the syringe is again checked for position and if necessary moved cranially so as to point the needle caudally; if this is not done contrast may spill up over the vocal cords. When injection is finished the needle is pulled out and positions 5 and 6 adopted. For the second side or for re-injection the needle is re-inserted through the same puncture site.

Complications and difficulties

1. The contrast may be injected into the tissues of the larynx. This should not happen with correct technique.

2. The puncture may appear satisfactory but no air can be drawn back. This indicates that the needle is plugged with soft tissues; it must be withdrawn, cleared and re-inserted. The chin *must* be in line with the sternum and must not be allowed to sag.

3. The vocal cords may be damaged. This is most unusual and can only occur if the needle is pointed cranially instead of caudally and then moved vigorously.

4. Bleeding may occur at the injection site. This can be troublesome but is usually no more than to give a slight staining of the contrast when it is coughed up. Really severe haematoma formation would necessitate tracheostomy. Late infection of a haematoma can cause abscess formation and even perichondritis.

5. Surgical emphysema is uncommon, unless one omits to instruct the patient to press on the puncture site when he coughs up the contrast.

The patient is warned:—

1. That his sputum may contain blood.
2. That his neck will be sore and tender for a period of some days if not for longer; that it is possible that he may feel a 'crackling' from air in the tissues.
3. That he may get a severe attack of coughing when the local anaesthetic wears off, and that he should press on the puncture site while he is coughing.
4. That he should not eat or drink for at least 3 hours.

OTHER METHODS OF BRONCHOGRAPHY

Variations on the crico-thyroid technique.—The crico-thyroid approach can be further extended by the use of the Seldinger technique for catheterisation of the trachea, (Craven, 1965). The crico-thyroid membrane is punctured in the manner already described, using either an ordinary crico-thyroid needle or a Riley or PE 205 Seldinger needle-cannula.[1] The tip of the needle or cannula is directed caudally and a 50–60 cm PE 160 Seldinger guide-wire is passed downwards into the trachea. The needle is withdrawn and a 30 cm. grey Kifa catheter with side-holes is passed over the guide and into the trachea.

A further variation on this approach is puncture of the trachea itself, either at the most easily palpable point, passing the needle between the tracheal rings, or in the relatively bloodless area immediately below the cricoid cartilage. This method (B. Kendall—personal communication), is best performed with a catheter-over-needle (e.g. Longdwel 6719). Ordinarily the catheter portion of the needle-catheter is used for injection of contrast but it will also allow the passage of a guide-wire so that selective catheterisation can be undertaken.

Oral Intubation methods.—The possible routes are:—

    (i) over-the-tongue intubation carried out 'blind'.

    (ii) intubation using an indirect laryngoscope.

Blind oral intubation is carried out in the same position as intubation through the nose, i.e. chin-out with the tongue pulled forward. The tube, which should be stiff with a good curve, is advanced in the midline over the back of the tongue and downwards toward the estimated position of the vocal cords. Thereafter the technique is similar to nasal intubation and is, in like fashion, by trial and error.

When the tube is in position it is strapped to the cheek and near the angle of the mouth. In small patients it must not be put in too far because of the risk of its going down the right main bronchus.

Indirect Bronchography and Laryngography.—When other methods fail or when it is desired to make the examination as simple as possible, the contrast may be introduced by dropping it over the back of the tongue between the vocal cords. This is also a useful approach for demonstration of the pharynx and larynx. Its disadvantage in bronchography is that the upper lobes are seldom well shown.

Anaesthesia is produced by the usual method except that to anaesthetise the vocal cords 3 ml. of 4% lignocaine are dropped on to the cords with the same curved cannula through which contrast is later instilled.

Contrast is introduced either by a long curved cannula or by a tube passed through the nose to the posterior naso-pharynx so that its tip is seen through the mouth. When the nasal tube is used it is put into the

---

[1] A relatively large needle is required before insertion of the grey Kifa tubing; otherwise it may be difficult to pass it through the crico-thyroid membrane.

nostril opposite the side to be examined, i.e. right nostril for left side and vice versa. The patient then leans slightly towards the side under examination. Similarly when the curved cannula is employed the patient leans to the appropriate side and the cannula is held a little to the opposite side, its tip vertically above the cords. Contrast is first injected in small amounts—3–4 ml.—as the patient breathes in, and positions 1, 2 and 3 are briefly adopted in between the individual injections. Finally the rest of the 20 ml. (the full amount is used) is rapidly injected during deep slow inspiration and the patient placed in positions 4, 5 and 6.

Laryngography.—When this method is used for outlining the larynx the contrast is dropped onto all sides of the pharynx and larynx as well as through the cords. Delineation is quickly checked by fluoroscopy (image intensifier) and P.A. (or A.P.), lateral and oblique films are taken, together with a P.A. and lateral while the patient performs the Valsalva manoeuvre. (The patient should be made to practise the Valsalva manoeuvre before the contrast is introduced.) 10 ml. of contrast are used.

BRONCHOGRAPHY UNDER GENERAL ANAESTHESIA

Examination of children.—In children under twelve we prefer to perform the examination under general anaesthesia. The method used is simple but needs the co-operation of an anaesthetist; two assistants are also required. A tunnel is a useful addition to the equipment especially for heavier children. Preliminary physiotherapy in the preceding week is most important and antibiotic therapy may also be of value. Apart from this the routine preparation and premedication for general anaesthesia are ordered. An anaesthetist is 'booked' for the examination.

Contrast.—Dionosil oily is the medium of choice. One of the assistants shakes the bottle of Dionosil—holding it between thumb and forefinger to avoid excessive warming—intermittently from the time the patient arrives.

Procedure

1. Before anaesthesia is induced the endotracheal tube to be used is chosen. *It should not be too long as its tip must lie above the carina.* A T-shaped Magill endotracheal connection is fitted into the outer end. Next a soft rubber tube through which the injection of contrast will be made is slipped into the endotracheal tube to make sure it slides through easily. When it just appears at the tip of the endotracheal tube it is marked with a loop of adhesive strapping at the other end (Fig. 22).

2. A plain A.P. film (without grid) is taken with the patient lying supine (on the tunnel if one is available). Most radiographers accustomed to working with children will be able to dispense with a preliminary lateral film, but this should not be omitted if there is any doubt as to the correct exposure or if the chest is asymmetrical.

3. Anaesthesia is now induced and the child intubated. The bronchi are then sucked out by the anaesthetist.

4. While this is being done the amount of contrast to be used is assessed and drawn up in a 10 ml. syringe. For each side we use roughly 1 ml. per year of age but give rather more if the child is large for his age. Plain film changes of bronchiectasis are an indication for 1–3 ml. extra per side. The procedure about to be carried out is explained to the assistants and in particular they are told what will be required of them for the *quick* and careful positioning necessary. One helps to position by moving the shoulders, the other by moving the hips. The latter assistant also covers the pelvis with lead rubber before exposures.

5. The child is now 'scolinised'. A film is put in the tunnel and the exposure is set for the first film. The lungs are vigorously inflated with oxygen for about 20–30 seconds. The side-tube is then disconnected from the Magill endotracheal connection and the occluding rubber cap removed. The operator slides the rubber tubing (now attached to a syringe

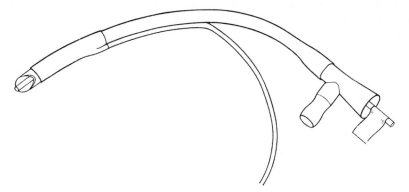

FIG. 22.—Bronchography under anaesthesia. Before anaesthesia the tube for injection is adjusted so that its tip is at the tip of the endotracheal tube. It is then marked with adhesive tape.

of contrast) down into the endotracheal tube to the correct level. The anaesthetist holds the head, one assistant the shoulders, the other the pelvis and under the operator's direction the child, now apnoeic, is turned onto the side under investigation with the foot end of the table tilted down some 25° when this is possible. All the contrast medium is injected in this position and after an injection of air to clear the tube the catheter is removed. The child is now turned half prone and then supine, a single careful squeeze on the anaesthetic bag helping to drive the contrast to the periphery of the lung fields. Filling is quickly checked on the monitor when this is available. During all this time the child is apnoeic. He is then placed in the right lateral position (or left posterior oblique if the examination is of the left side only). The anaesthetist reconnects his tubing and closes the open end of the Magill connection; he allows the anaesthetic bag to fill with oxygen and when it is full he squeezes it to inflate the chest. At full inflation he pauses and calls 'take', allowing expiration to take place when the exposure has been made. The child, again apnoeic, is now turned into the A.P. position, re-inflated and a further exposure is made in a similar manner. (No inflation of the lungs is made between exposures.) With each successive inflation the contrast is driven further distally, although if the correct amount has been chosen it will not go too far. The child lies on the side under examination until the films have been viewed. 'Repeats' are seldom necessary except for underfilling of bronchiectatic areas. Before proceeding to examine the second side the first, i.e. right, is sucked out by the anaesthetist.

It is in fact wise to examine the opposite side in any case showing significant bronchiectasis, even when this has not been requested. This will provide a fuller picture of the child's condition, making a more adequate basis for the planning of treatment whether surgical or medical.

This technique gives a picture of the bronchi in full inflation and although it separates them out very elegantly the appearances are somewhat artificial. It may be preferred to take the films in relaxation immediately after inflation. In any case where there is doubt about the presence of bronchiectasis a 'relaxed' film is helpful in showing the bronchi in a more normal state.

When the examination is over the anaesthetist sucks out the bronchi and the child is returned to the ward. A film after 24 hours is useful because areas of collapse or residual contrast may be shown which will indicate the appropriate positioning for physiotherapy. The latter is continued for 3–4 days after the procedure, beginning as soon as the child is sufficiently awake to co-operate.

Examination in adults.—A similar technique is suitable for examination of adult patients and is particularly appropriate for use in conjunction with bronchoscopy (Le Roux and Duncan, 1964). When the bronchoscopy has been completed the lungs are inflated with oxygen and the trachea is intubated with a cuffed endotracheal tube, the cuff being inflated. The patient is turned onto the right side and a cassette is inserted beneath him. The X-ray tube is centred, a catheter is passed down the endotracheal tube and contrast is injected. The tube is withdrawn, the rubber cap of the Magill connection is replaced and the lungs are forcefully inflated for five breaths. The first exposure is made at the completion of the fifth inflation and the patient is then turned onto his back for the A.P. film which is also taken in full inflation. The left side is similarly examined with an oblique and an A.P. film. At the end of the examination the contrast is removed from the bronchi, using a sucker.

Either Dionosil Oily or Hytrast are suitable for this method, giving 15–20 ml. to each side in the normal adult. The addition of $\frac{1}{2}$–1 ml. of water to the Hytrast is essential to prevent bronchial obstruction and improve peripheral filling.

Difficulties and complications

The difficulties in achieving satisfactory injection of contrast have been discussed earlier. It is essential to achieve the best possible anaesthesia before attempting intubation or injection of contrast since coughing can spoil an otherwise technically adequate bronchogram. Failure of contrast to enter a particular part of the lung may be attributable to insufficient injection of contrast but may also be due to disease in the area concerned. Contrary to what was thought in the past the movement of contrast medium is largely due to the passage of air along the bronchi and regions of the lung with airways obstruction do not show good filling with contrast medium. This is particularly true of areas of emphysema and may also be true of areas of bronchiectasis where the alveoli distally are almost functionless. In these circumstances prolonged posturing, if necessary with selective injection of contrast medium, may be the only means of producing adequate filling and selective injection may also be needed for areas where there is obstruction. A description of this approach is beyond the scope of this book.

The use of excessive amounts of contrast medium may produce areas of distal collapse or of 'contrast pneumonia'. When a tube of adequate length is still present in the trachea it may be possible to undertake some form of bronchial suction, otherwise active physiotherapy is the most useful means of re-expanding an area of collapse. Most patients experience mild discomfort and some fever following injection of contrast medium of whatever sort but this seldom lasts for more than 24 hours. A chest film taken at this stage will often show the adequacy of clearing and may indicate the appropriate position for continued physiotherapy.

References

AMPLATZ, K. and HAUT, G., (1970). *Radiology*, 95, 439.
CRAVEN, J. D., (1965). *Br. J. Radiol.*, **38**, 395.
Le ROUX, B. T. and DUNCAN, J. G., (1964). *Thorax*, **19**, 37.
PRIDIE, R. B., (1966). *Br. J. Radiol.*, **39**, 73.

APPENDIX

Preparation

1. Arrange preparatory physiotherapy if needed.

2. *O.P. Instructions*

    The X-ray examination is at .... on ....

    1. Have nothing to eat or drink after .... a.m./p.m. (usually 4 hours before the appointment)

    2. Come to the X-ray department at ....

3. *Ward Instructions*

    The bronchogram is at .... on ....

    1. The patient is to have nothing to eat or drink after .... a.m./p.m. (from 4 hours before the appointment)

    Give 1 Xylocaine lozenge at .... a.m./p.m.

After-care

1. Postural coughing and percussion is arranged with the physiotherapy department in appropriate cases.

F

2. The patient is told not to eat or drink for 3 hours after leaving the department. The time when he may eat is written on a piece of paper and handed to the patient. This is important as the stress of the examination may make the patient forgetful; a carbon copy may be kept by the X-ray department.

BRONCHOGRAM

Trolley settings

Crico-thyroid route

Sterile—upper shelf

1 towel; gauze and wool swabs—plentiful.
1 gallipot for iodine; 1 gallipot for spirit.
2 medium receivers.
1 × 2 ml. syringe with subcutaneous and intramuscular needles.
1 Lipiodol syringe and needles *or* 20 ml. syringe and Longdwel 6719.
Needle guard if used.
or Riley needle, 50 cm. PE 160 guidewire and 30 cm. grey Kifa catheter.

Lower shelf

Bowl (with mild antiseptic) for expectoration.
Bottles of 2% and 4% lignocaine topical. Ampoule of saline for dilution.
Vial lignocaine 1% for injection.
Bottles contrast medium (Dionosil Oily).
Masks.
Nobecutane spray or Elastoplast airstrips.

Intubation methods

Sterile—upper shelf (It is not practicable to maintain full asepsis but the trolley is laid out with all items sterile.)
Gauze swabs—plentiful.
Assortment of Jacques catheters.
Short Seldinger guide-wires to provide necessary curve.
1 spigot.
Spray for lignocaine e.g. the Xylocaine spray—(Astra-Hewlett, Ltd., Watford) *or* Rogers' crystal spray (1 Beaumont Street, London, W.1).
1 × 20 ml. syringe (for contrast).

Lower shelf

Bowl (with mild antiseptic) for expectoration.
Bottles of 4% and 2% lignocaine or 4% prilocaine with ampoules of saline for dilution for surface anaesthesia.
Torch or head mirror.
Masks (the operator wears a mask since the patient may cough in his face).
Jar of Xylocaine gel. Orange sticks.
Strips of adhesive plaster (for taping the tube to the cheek).
(Laryngoscope—Mackintosh's.)
EMERGENCY EQUIPMENT.

CHAPTER 20

DIAGNOSTIC PNEUMOPERITONEUM

THE gases used for this examination may be divided into two groups: those which are slowly absorbed—air and oxygen—and those which are rapidly absorbed—carbon dioxide and nitrous oxide. The latter gases are so soluble in the blood stream that if they are accidentally injected there is little or no risk of gas embolism. This has led some authors (e.g. Lumsden and Truelove, 1957) to use them exclusively but the disadvantage of these gases is that there is only a short time available for taking films since most of the injection will have disappeared within 30–40 minutes. It is possible, however, to gain the advantages of both types of gas by inducing the pneumoperitoneum (P.P.) with nitrous oxide or carbon dioxide until the disappearance of liver dullness shows that the P.P. is safely established and the needle is correctly positioned. This usually requires about 500 ml. of gas. Thereafter, if plenty of time is required for taking films, the filling is completed with air or oxygen.

Contra-indications

1. Severe emphysema or heart disease.
2. Multiple abdominal incisions make adhesions likely and demand more than usual care and the use of CO_2 or N_2O.
3. Ascites is not a contra-indication but the fluid is first drawn off, using a lower puncture than for routine pneumoperitoneum.

Ward/O.P. Preparation

1. *Premedication.*—Valium 5–10 mg. is given intramuscularly 45 minutes before the procedure both to in-patients and to nervous out-patients. Scopolamine 0·4 mg. is given intramuscularly 45 minutes before the induction. This is intended to reduce the risk of reflex vagal over-action.
2. *Bowels.*—Routine preparation as for urography is given, both to clear the radiographic field and reduce the risk of puncture of a gas distended colon. The rectum and sigmoid must be *completely* empty for pelvic pneumoperitoneum and for this reason patients are better admitted to the ward before the examination to undergo in-patient bowel preparation (p. 5).
3. *Bladder.*—This is *emptied immediately before* the examination—again this is of greatest importance in pneumo-pelvigraphy.
4. The patient has nothing to eat or drink for at least 4 hours before examination.

Summary of the Procedure

1. Plain film of abdomen for adequate bowel clearing and bladder emptying.
2. Palpate the abdomen; percuss out and mark the edge of liver dullness.
3. Prepare the skin of the abdomen including the area of the liver.
4. Select puncture site and infiltrate skin and abdominal wall down to the peritoneum with local anaesthetic.
5. Puncture abdominal wall with appropriate needle.
6. Test for correct position of needle.
7. Introduce N_2O at least until liver dullness disappears.
8. Complete filling with air or oxygen if desired; a total of 1,000–1,500 ml. is given and 2,000 ml. before pneumo-pelvigraphy.
9. Position and take appropriate films.

Equipment

The needle for induction

1. One needle commonly used for this purpose is the Saugmann needle. Its only advantage is that its stilet is longer than the needle, making it possible to probe the region beyond the needle tip. In very fat patients the Saugmann needle may be found too short and a needle up to 4 or 5 inches long is occasionally required.
2. Other needles of a similar gauge serve equally well, e.g. 18 gauge arterial cannula (M.D.). A short flexible wire similar to the Seldinger guide-wire may be used to probe beyond the tip of the cannula. An arteriogram connection is required to join the needle to the syringe—with the Saugmann needle rubber tubing is used.

Source of gas.—This may take many forms. We have found it convenient to use an anaesthetic rebreathing bag charged with the appropriate gas (see Appendix). It can be used directly or via a pneumothorax apparatus.

Department Preliminaries

1. A plain film of the abdomen is taken. If the rectum and colon are not empty before pneumo-pelvigraphy the procedure should be postponed. For other areas this is less important, but any areas of colonic distension are noted so that they may be avoided in subsequent puncture.
2. The examination is explained to the patient, with a warning—where air or oxygen is to be used—that there will be a feeling of abdominal distension for a few days. He is told also that he is likely to feel pain in the shoulders.
3. The abdomen is palpated and percussed carefully for possible masses and the edge of the normal liver dullness is found and marked. Ab-

dominal scars are noted so that they may be avoided when puncturing the abdomen. In patients with abdominal distension the presence of shifting dullness is sought. (Ascitic fluid, when present must first be drawn off. This involves puncturing in the right or left iliac fossa just outside McBurney's point, which is half-way between the anterior superior iliac spine and the umbilicus. A wider needle—i.e. 16 or 17 gauge—is used or a suitable trocar and cannula. When most of the fluid has been removed, which usually involves appropriate tilting of the patient into the foot-down position, the needle is connected to the pneumothorax machine and gas is injected. A larger total, at least 1,500 ml., is advisable.)

Induction of the Pneumoperitoneum

1. The operator, wearing a mask, scrubs up and prepares the abdomen, using iodine in the region of intended puncture and chlorhexidine/spirit over a wide area to make possible subsequent percussion of the liver region. A sterile towel is laid over the pelvis and lower abdomen.

2. The usual sites for puncture are (a) just lateral to the left rectus muscle from about $1\frac{1}{2}''$ above to $1\frac{1}{2}''$ below the umbilical level, (b) in the midline, just below the umbilicus. However, scars or masses may make puncture outside these regions necessary or the right side may be used. The skin and subcutaneous tissues are anaesthetised with 1% lignocaine and the skin is punctured with a cutting needle. A long intramuscular needle is attached to a 10 ml. syringe of 1% lignocaine for infiltration of the abdominal wall. To make this infiltration easier to control, the patient lifts his legs thus tensing the abdominal muscles or the linea alba so that the needle can be felt to pierce them. In thin patients and those of average build anaesthesia is produced down to the peritoneum, but in fat patients a longer needle is sometimes needed to complete the anaesthesia.

When the local anaesthetic has been injected the needle for induction is tested for patency with a stilet, and it is then introduced through the skin. The patient again tenses the abdominal wall and the needle is advanced. The piercing of the transversalis fascia or linea alba and the peritoneum may each be felt as a distinct 'give', sometimes associated with a prick of pain. Only when the peritoneum is thought to have been penetrated is the patient allowed to relax the abdominal wall. (For the first few inductions the needle is advanced rather deliberately with a conscious effort to recognise the components of the abdominal wall, and using a to-and-fro rotation to make the advance more controlled. With experience it will usually become possible to thrust the needle straight into the appropriate level). For any subsequent advance of the needle the abdomen must be tensed again since in thin patients it is possible to reach the posterior abdominal viscera if the abdomen is relaxed.

Once the needle is thought to be in position the patient relaxes and the following confirmatory manoeuvres are carried out:—

1. The stilet is passed down the needle. The Saugmann stilet is longer than the needle and so can probe the region beyond the tip. As already suggested a suitable length of flexible wire may be used with other needles. Firm resistance to this probing indicates that the needle is in a viscus or in the abdominal wall—it is usually easy to judge which is the more likely. Coils of intestine offer a gentle resistance which feels quite different.

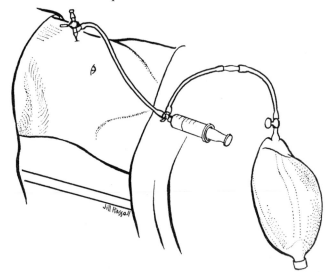

FIG. 23.—Pneumoperitoneum. Simple assembly for injecting N_2O, CO_2 or O_2. To inject air the tubing and anaesthetic bag are disconnected from the side-arm of the two-way tap.

2. The patient takes a deep breath. If a solid viscus has been pierced this will cause a big swing of the needle and it must be withdrawn. Normally only a slight rocking occurs.
3. The needle is joined to the two-way tap and 50 ml. syringe as shown in Fig. 23. The plunger of the syringe is drawn back to confirm that neither a vessel nor a gas-containing viscus has been entered.
4. 20–30 ml. of N_2O from the syringe should inject easily with a minimum of resistance and with no pain (when using the Saugmann needle the tap on it is closed during injection of gas). Distinct resistance or pain are both signals to stop injecting, pass the stilet down the needle and adjust if necessary. When a pneumothorax apparatus is employed the pressure is measured after injection of 50 ml. of gas. If the needle lies in the peritoneal cavity the gas dissipates easily and the pressure will remain close to zero. A rise in pressure indicates extra-peritoneal injection, usually into the abdominal wall.

When the satisfactory positioning of the needle is confirmed, injection of the full quantity of gas may begin either by means of the 50 ml. syringe but preferably using the Maxwell box since this makes injection considerably quicker. Provided all goes smoothly at least 500–600 ml. of N_2O are injected (before pelvic pneumography 1,000 ml.). At this stage liver dullness should have disappeared confirming establishment of the P.P. It is then safe to instil air or oxygen of which 500–1,000 ml. are introduced for general diagnostic purposes and 1,500 ml. for demonstration of the stomach or for pneumopelvigraphy. Before the latter examination the patient may be tilted head-down during injection of gas for greater comfort. Although no pain should be felt at the injection site an uncomfortable abdominal tightness and shoulder pain referred from the diaphragm are common. N_2O is less liable to cause discomfort and if one is confident that all necessary films can be obtained in 30 minutes or less, N_2O alone may be injected.

When filling is complete the needle is removed and the puncture site sealed with collodion. The method of examination varies thereafter and a description of technique for various areas will be given.

It may be pointed out that some authors (Schulz and Rosen, 1961; Frimann-Dahl and Traetteberg, 1962) advocate aspiration of the gas when the examination is finished. This is certainly logical if large volumes of poorly absorbed gas are used but we have no personal experience of this manoeuvre.

General Diagnostic Pneumoperitoneum

Pneumoperitoneum may be used to outline upper abdominal masses, to show enlargement or irregularity of the liver or spleen and to demonstrate adhesions to the anterior abdominal wall. Each case presents its own particular problems and some thought is required to plan the most suitable views. For most cases the positions recommended by Lumsden and Truelove (1957) are those of greatest value. These are:—

1. *Prone film with overcouch tube using the Potter-Bucky grid.*—When only one side is to be examined the patient lies on the opposite side for 1–2 minutes, breathing slowly and deeply. The gas rises into the region on which interest centres; the patient is then turned into the prone position for the film.

To distribute the gas evenly on both sides the patient is brought upright on the tilting table, turned to face the table and tilted down into the prone position.

The film used should be large enough to include the diaphragm and the side walls of the abdomen. It is not usually necessary to include the region below the iliac crests.

2. *Supine lateral with horizontal ray and without grid.*—This view shows the posterior aspect of the abdominal wall and the anterior surface of the liver. Elevation of one side or the other can be employed to bring masses or adhesions in the region of abdominal scars onto the 'skyline'.

3. *Erect P.A. or A.P. without grid.*—This is of value in showing the diaphragm (see below). When it is used for upper abdominal masses we prefer to use a grid.

4. *Other views include.*—

(i) Oblique films in the prone or erect position; these may be taken under screen control if desired.

(ii) Lateral decubitus films with horizontal ray.

Tomography may be used in cases in whom air or oxygen have been injected. However, its value is somewhat limited if the available tomographic table cannot be tilted from the horizontal position. The positioning for tomography is decided on after viewing the initial films. As with presacral pneumography 'labelling' of various organs may be indicated, e.g. cholecystogram, urography, barium swallow.

Pneumoperitoneum to show the diaphragm.—P.P. may be used to establish the position of the diaphragm and show any deficiencies in it or adhesions to it. A problem which is sometimes encountered is the apparent elevation of one hemi-diaphragm, particularly the right. In addition to phrenic paralysis this may be due to lesions above the diaphragm, e.g. infra-pulmonary effusions, hernias through it or enlarged viscera below it. In such cases P.P. can be helpful and the lateral film is sometimes the most revealing. It may be necessary to take a lateral with the patient leaning forward to show the diaphragm posteriorly. It should be remembered that gas may be slow to penetrate small gaps, e.g. around a liver which has herniated through the diaphragm. For this reason:—

(ii) air as well as N_2O should be used.

(ii) films taken after the patient has been upright for 1–2 hours or more may be of value if initial films fail to define the diaphragm clearly.

Pneumoperitoneum to show the stomach wall.—The filling defect in the gastric fundus which might be due to an intrinsic lesion or to extrinsic pressure is a not uncommon problem which can be very difficult to resolve. Use of pneumoperitoneum may make it possible to distinguish between intrinsic and extrinsic lesions, to prove that an intrinsic filling defect is due to coarse rugae or to show the extent of a neoplasm more completely than a simple barium meal.

The P.P. is induced in the usual way and 1,500–2,000 ml. of gas is instilled since, as Frimann-Dahl and Traetteberg (1962) point out, it is important to fill the lesser sac. After induction the table should be tilted into the semi-erect position, the patient lying first prone and then supine to encourage movement of gas into the lesser sac.

To outline the inner aspect of the stomach barium may be given. Alternatively the stomach may be distended with gas either by giving an effervescent powder (Frimann-Dahl and Traetteberg, 1962), *or* by introducing gas through a stomach tube (Lumsden and Truelove, 1957). The advantage of gas distension is that the stomach wall is stretched, smoothing out irregularities due to coarse rugae or to peristalis. However, eructation may make repeated distension necessary.

Barium, when used, is swallowed in the supine position in order to coat the gastric fundus. The patient is then brought upright and films are taken erect in varying degrees of obliquity and with the patient tilted back or leaning forward to bring the gas to particular regions. Overcouch prone-oblique films may also be of value.

When gas distension of the stomach is used films are taken either with the overcouch tube or during fluoroscopy. The semi-erect supine or erect positions are the most useful; the degree of obliquity needed varies. Frimann-Dahl and Traetteberg (1962) obtain tomographs in the semi-erect or near-erect positions. Their results are most satisfactory but as relatively few hospitals possess tomographic tables which can be used when tilted, this method has a somewhat restricted application.

Pneumopelvigraphy (Gynaecography).—Here the aim is to show the pelvic viscera, chiefly the uterus and ovaries, by filling the bowl of the pelvis with gas. Except when the bladder is to be examined the bowels and bladder must be completely empty. (If the plain film shows bladder distension the patient is catheterised.) Some form of myelographic harness or support is needed to hold the patient. After the P.P. has been induced this equipment is prepared and the patient lies prone. She is then tilted 50° head down, taking deep breaths as she does so.

The film is taken with the tube angled 10° from the vertical toward the feet (Fig. 24). The ray is centred on the coccyx and should pass on a line about 2″ above the greater trochanters. The film is put in the Bucky tray and placed somewhat cranially to allow for the angling of the ray. It is as well to use a large cassette since the area concerned can easily be 'thrown off' a small film.

Should the P.A. film fail to show the uterus and ovaries clearly, oblique views may be taken with left or right side raised.

Complications and Difficulties

These are fully reviewed by Stein (1951). Among the commoner ones are:—

1. *Pain in the shoulder.*—This is said to be due to stretching of sub-phrenic adhesions. It is certainly due to referral of pain from the diaphragm and the head-down position may relieve it.

2. *Interstitial emphysema.*—This is usually of the abdominal wall and scrotum but is occasionally retroperitoneal. It may be unpleasant but is seldom harmful. If N_2O is used it will not interfere with the films and, after the needle has been repositioned, induction may be resumed and the examination continued.

3. *Haematoma of the abdominal wall.*—Bleeding is unlikely to occur if

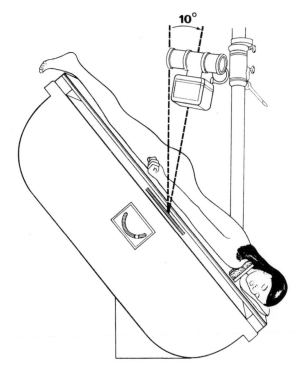

FIG. 24.—Pneumopelvigraphy. Position of the patient for radiography.

puncture is kept to the areas mentioned and away from the region of the superficial epigastric arteries.

4. *Penetration of a hollow viscus.*—A well recognised hazard, but although it spoils the examination it rarely causes peritonitis.

5. *Penetration of a solid viscus.*—This is usually the result of failure to take sufficient care in initial palpation. It is not invariably harmful but occasionally troublesome haemorrhage may result. If air is injected gas-embolism may ensue.

6. *Gas embolism.*—This is the most important complication and its treatment—outlined in Chapter III—should be learnt before undertaking P.P. It is recognised by:—

 i. Sudden collapse of the patient during instillation of gas.
 ii. Chest pain.
 iii. A 'churning' noise in the chest.

Induction with N_2O reduces this risk to a minimum.

7. *Vagal over-action.*—It is well known that irritation of serous membranes can produce reflex vagal over-activity. This may result in a simple vaso-vagal attack but can very rarely give rise to cardiac arrest. The risk is much lower than with stimulation of the pleura but we consider it

advisable to give scopolamine before the examination so as to reduce this risk to a minimum.

8. *Scrotal pneumatocoele* in inguinal hernia.

References and further reading

Pneumoperitoneum

FRIMANN-DAHL, J., and TRAETTEBERG, K. (1962). *Br. J. Radiol.*, **35**, 249.
LUMSDEN, K., and TRUELOVE, S. C. (1957). *Br. J. Radiol.*, **30**, 516.
STEIN, H. F. (1951). *Am. Rev. Tuberc.*, **64**, 645.

Gynaecography (*Pneumo-pelvigraphy*)

SCHULZ, E., and ROSEN, S. W. (1961). *Am. J. Roentg.*, **86**, 866.

APPENDIX

Ward/O.P. Preparation

1. Nothing to eat or drink for 6 hours before.
2. Bowel preparation: as for I.V.U.; before pneumopelvigraphy the 'in-patient' preparation is given.
3. The bladder is emptied immediately before the patient leaves the ward and again in the X-ray department if there is any delay in induction.

Trolley setting

Sterile—upper shelf

3 sterile towels; gauze and wool swabs.
2 gallipots, one for chlorhexidine/spirit, one for iodine.
1 small receiver with a 10 ml. syringe and nos. 1 and 17 needles.
50 ml. syringe.
6 cm., 16 gauge needle or ascites trocar and cannula (in case of ascites).
Gloves. Paper hand towel.
no. 11 disposable blade (for dividing skin).
either Saugmann needle and 2 × 1 foot lengths of rubber tubing of suitable gauge.
　　　Two-way tap with front and side arms for rubber tubing—Luer fitting.
　　　Glass connection with sterile wool on it and 2 foot length of wider rubber tubing for connection to N_2O bag or Maxwell Box.
or　18G Arterial cannula (M.D.) or lumbar needle and
　　　Arteriogram connection and
　　　1 foot length of rubber tubing and glass connection as above and
　　　2 foot length of wider tubing.
　　　Two-way tap with Luer male fitting on front arm; side arm for rubber tubing.
　　　Short, flexible wire stilet for use with arterial cannula (M.D.).

Lower shelf

Lignocaine 1 %, 2 vials of at least 20 ml.
Bottles of iodine and chlorhexidine/spirit.
Collodion or equivalent.
Masks.
Small dressing strip.
EMERGENCY EQUIPMENT.

Also required

Source of N_2O (and O_2 if necessary)

We have used a 1, 1½ or 2 gallon anaesthetic bag of the double ended type used with a Waters's canister. A rubber tube with a small nylon tap is fitted into the narrow end and the wide end is sealed with a rubber bung. The bag can be filled with gas via the tap which is then closed (Fig. 23).

Maxwell Box (G.U.) or other pneumothorax apparatus.
Oxygen cylinder, reducing valve and flow meter.

CHAPTER 21

PRESACRAL PNEUMOGRAPHY

THIS examination can be of value in delineating retroperitoneal structures —para-aortic glands, kidneys, pancreas and the adrenal glands. Unfortunately the risk of gas-embolism is always present and the only safe gases—carbon-dioxide and nitrous oxide—are so rapidly absorbed that considerable speed is required to obtain satisfactory films before the gas has disappeared. For this reason most workers continue to employ air or oxygen. The technique used must, therefore, contain all possible safety measures and these will be noted in the following account. In particular we believe that the use of nitrous oxide or carbon dioxide in the initial stages of the examination until the gas has been visualised passing into the perirenal area, offers a considerable addition to the safety of the procedure.

The investigation of the adrenal glands

The indications for presacral pneumography are diminishing. It is relatively inaccurate in the evaluation of smaller tumours and particularly in assessing the adrenals in Cushing's syndrome, where the abnormal fat may adhere to the glands, concealing their true size. Larger tumours of all types are often demonstrable by means of high dosage urography (e.g. 100 ml. Conray 420) and immediate tomography, whilst smaller tumours are best shown by adrenal venography (Starer, 1965; Mikaelsson, 1967; Bookstein, Conn & Reuter, 1968). The suggested routine for each type of suspected abnormality is as follows:—

1. *Cushing's syndrome.*—High dosage urography and tomography in all cases. If a firm diagnosis of tumour can be made no further investigation is required. If nephrotomography is negative or equivocal presacral pneumography is indicated, followed, if necessary by adrenal venography or arteriography.

2. *Phaeochromocytoma.*—Preliminary control of catecholamine excretion is instituted as described in Chapter 3 and the precautions there outlined are taken. High dose nephrotomography is performed and if this is negative presacral pneumography is carried out. Further investigation by selective adrenal arteriography or venography may be required if pneumography is inconclusive.

3. *Conn's syndrome.*—Retrograde adrenal venography is the investigation of choice once a firm biochemical diagnosis has been made. Manipulation should be as gentle as possible and contrast injection should be made slowly and watched on the screen to avoid extravasation or over-injection. After injection of contrast the catheter should be withdrawn a little to prevent overfilling of the gland by the flushing solution.

Contra-indications

1. Emphysema, heart disease.
2. Perianal sepsis.
3. Retroperitoneal fibrosis or infiltrating retroperitoneal tumours.

Special points when booking the examination.—It is preferable to carry out the induction on a tilting table and one should be able to take tomographic cuts as well. With some methods screening will be needed.

Ward preparation

1. In-patient bowel preparation (p. 5).
2. Fast for at least 4 hours.
3. Shave natal cleft and perianal area.
4. Empty bladder before coming to the department.
5. Omnopon 20 mg. for an average (70 Kg.) adult, is given three-quarters of an hour previously.

Department preliminaries

1. *Plain film of abdomen*
   (a) For the adequacy of bowel clearance.
   (b) For exposure check. The correct exposure is 'pale' since the gas when in the tissues leads to a considerable darkening of the film.
   It is preferable for the bowels to be well cleared since even tomograms may be affected by heavy faecal shadows. A preliminary A.P. tomographic cut is made at the estimated level of the kidneys.
2. The procedure is explained to the patient warning him to expect a feeling of some tightness and 'crackling' in his back, chest and neck. This applies chiefly when air or oxygen are used.
3. If the patient seems unduly nervous, 5–10 mg. of Valium may be given intravenously, but heavy sedation is undesirable as the patient may later need to sit up.

Contrast.—The choice of gas to be used is a matter of opinion. Carbon dioxide or nitrous oxide are undoubtedly safest, carrying almost no risk of gas embolism. Their use demands swift technique in all respects and since few departments perform more than occasional pneumograms this may be difficult to achieve. It is therefore advisable to use N_2O in conjunction with a catheter or cannula inserted into the retro-rectal tissues since this allows for repeated injections. It may also be used, when screening is employed, as an initial injection which may be followed by air or oxygen when screening shows that the gas is passing satisfactorily into the peri-renal space.

Air and oxygen are equally dangerous in respect of gas embolism but there *is* a case for using oxygen because it will be absorbed more rapidly than air and so cause less discomfort on the succeeding day.

Summary of the procedure

1. Plain film of abdomen; preliminary tomographic cut.
2. If the plain film is satisfactory place the patient in the left lateral position.
3. Clean peri-anal region; infiltrate with local anaesthetic as described below.
4. Introduce the 'needle' directing it towards the lower sacrum and infiltrating liberally with dilute lignocaine. The 'needle' being satisfactorily positioned, turn the patient into the semi-prone position and lower the foot of the table.
5. Inject 400–500 ml. of N_2O and screen the uppermost kidney to confirm that the gas is passing normally round the kidney. Inject 800–1200 ml. of oxygen, turn the patient half onto the other side and inject a further 800–1200 ml.
6. Sit the patient upright for 3–5 minutes.
7. Plain film of abdomen.
8. Oblique films and tomographs as required, preferably in the erect position.

Procedure

1. The patient lies on his left side with the knees drawn well up. It is helpful if an assistant draws the uppermost buttock upward, so opening the natal cleft.
2. The operator scrubs up, wearing a mask and gloves. The buttocks and natal cleft are cleaned with chlorhexidine/spirit as near to the anus as possible. Sterile towels are draped around the area and on the table top. The anus itself is too sensitive for spirit or iodine; Betadine is used to clean it, with several changes of swab. During the insertion of the needle the anus is covered with a swab.
3. Palpation is carried down the posterior aspect of the sacrum, over the sacro-coccygeal angle and down to the tip of the coccyx. This gives an idea of sacral curve and the angulation in the sacro-coccygeal region.

Midway between the tip of the coccyx and the anus is the fibrous ano-coccygeal body. A skin wheal of local anaesthetic is raised to one side of the midline between this point and the coccyx. An intra-muscular needle is used to infiltrate the subcutaneous and deeper tissues with lignocaine 1 %. This needle is directed backwards and towards the midline into the hollow of the sacrum. Drawing back on the plunger with each advance of the needle confirms that the needle is neither in a vessel nor in the rectum. Once the deeper tissues have been infiltrated a longer (3–3½", 8–9 cm.) needle is used. A variety of needles and cannulae may be employed, some of which will be discussed below. Although for convenience we will refer to the use of a 'needle' in the ensuing descriptions it is our belief that the actual injection of gas should not be made through a needle. This is because the use of a cannula or catheter will diminish the risk of inadvertent puncture of a vessel and so reduce the chances of producing gas embolism. Among the possible approaches are:—

1. To use a femoral arteriogram or lumbar puncture needle (if the latter, a simple type—e.g. Barker's or Pannett's—is preferable since side arms are awkward in the natal cleft).
2. To use a needle-cannula, i.e. a needle within a cannula, withdrawing the needle from the centre when it has reached the correct position. The advantage of this is that once the cannula is in position small shifts of its tip will not pierce vessels—always a possibility with a needle.
3. To use a needle-catheter, e.g. Longdwel 6719. Once the needle has reached the correct position and saline has been injected to expand the presacral space, the catheter is grasped firmly and pressed inwards while the needle is withdrawn. The catheter assembly is firmly fastened to the skin with adhesive tape. This is the best method.

The actual introduction of the needle is along the line previously described (Fig. 25), i.e. directed to pass just anterior to the lower sacral margin and into the sacral hollow in the midline. (It should be appreciated that in some patients the relationship between anus and coccyx is such that it is not possible to direct the needle into the hollow of the sacrum. The aim then should be to place the needle tip just anterior to the lower sacrum (Fig. 25d)). The needle is attached to a 20 ml. syringe of $\frac{1}{2}\%$ lignocaine which is used to anaesthetise and to expand the retrorectal tissues. Before inserting the needle the skin is nicked with a small scalpel (this is particularly necessary when a catheter is to be employed). The needle is advanced with its bevel directed forward to reduce the risk of impaling the rectum. With each advance the plunger is withdrawn and then 4–5 ml. of dilute lignocaine injected. As the needle passes more deeply it may strike the sacrum; this calls for a slight withdrawal, re-direction more anteriorly and a further advance. It is preferable at first to direct the needle too far back rather than too far forward. In fat patients, however, it is possible to direct the needle *behind* the sacrum and it will then be felt to move under the skin of the natal cleft. A satisfactorily placed needle cannot be felt in this way and when twisted to bring its tip backward it is held by the resistance of the sacrum.[1]

---

[1] Some workers have recommended advancing the needle with the left hand while a gloved finger in the rectum is used to confirm, by rolling the mucosa over the needle tip, that the rectum has not been caught by the needle. Others advance the needle and then do a rectal examination. The disadvantages are:—
1. Maintenance of asepsis becomes more difficult.
2. Those patients whose obesity makes it most difficult to assess the correct direction of the needle are those in whom rectal examination is least helpful (Sowerbutts, 1959).
3. Direction of the needle is much more easily controlled using both hands.
4. An assistant is needed to inject the gas.
In fact, rectal examination is unnecessary if plenty of dilute local anaesthetic is used and the needle is directed well posteriorly so that if anything it will strike the sacrum and can then be advanced just anterior to it.

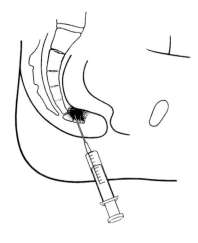

a.

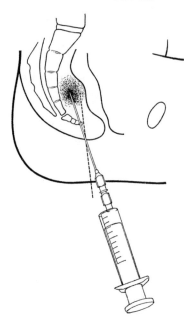

b.

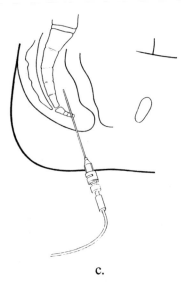

c.

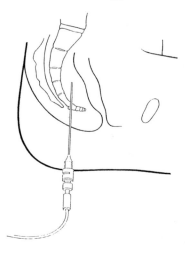

d.

Fig. 25.—Presacral pneumography. The insertion of the needle-catheter. *a.* Initial local anaesthesia. *b.* Deeper infiltration with dilute local anaesthetic to displace the rectum forward. Although the needle-catheter is directed into the hollow of the sacrum it is not uncommon for it to swing forward as shown when the syringe is detached, *c.* The catheter positioned anterior to the sacrum. *d.* In patients who are fat or who have a marked sacral curve with a short distance between anus and coccyx, it is not possible to direct the needle into the hollow of the sacrum. The aim is then to pass the catheter just anterior to the lower sacral margin.

The needle is advanced as described until it passes through the pelvic floor. This is indicated by a sudden lowering of resistance to injection. The position is confirmed by screening in the lateral position when the needle should lie as in Fig. 25, though its point may be somewhat more anteriorly owing to the tension of the tissues. When it has been shown by drawing back on the plunger to be neither in a vessel (no blood) nor in the rectum (no gas), 20 ml. of saline are injected further to expand the presacral space and make it relatively more avascular. If the needle-cannula or needle-catheter are used the needle is removed and injection of gas can take place through the cannula or catheter. A little gas is injected with the patient lateral and the satisfactory position checked by screening. While the operator holds the catheter or cannula firmly in position the patient is then told to turn into the semi-prone position, bringing the left arm underneath and straightening the hips. The foot of the table is lowered giving a tilt of about 15 degrees. The catheter is then taped to the skin.

Injection of gas.—The needle, cannula or catheter is now joined by an arteriographic connection to a two-way tap on a 50 ml. syringe. The side arm of the tap may be open (when air is used) or may be joined by rubber tubing to a source of O_2 or N_2O or to a pneumothorax apparatus drawing in O_2 or N_2O. The Maxwell box is convenient for this purpose.

Gas is drawn into the syringe and then injected. There should be only a slight sense of resistance when the gas is going in. A resistance to injection suggests obstruction of the needle and demands:—

1. Clearing of the needle with a stilet.
2. Small shifts in its position and angulation.

Once the gas is running in satisfactorily it may be injected fairly rapidly —about 50 ml. in 10 seconds. After each syringeful the total injected is called out and recorded and where a needle is used the plunger is pulled back to confirm that it is not in a vessel. If the pneumothorax apparatus is to be employed the syringe is only used until it has confirmed that the gas runs in easily. Some pain is often experienced during injection of the gas. To begin with nitrous oxide is injected—about 400–500 ml.—until it is clear on fluoroscopy that it is passing normally around the right kidney. It is then safe to inject a more slowly absorbed gas i.e. oxygen or air. The volume injected varies with the size of the patient. 600–800 ml. on each side is adequate for thin patients; the obese need 1,000–1,200 ml. for each side. With N_2O alone rather more is required because of the rapid absorption. When the amount for the first side has been injected the patient's legs are carefully straightened, the operator holding the 'needle' steady. The patient is rolled onto his face and then halfway onto his right side. A further injection of gas may then be made.

It is at this stage that symptoms from gas-embolism are particularly likely to occur since the protective effect of the left lateral position is removed. The operator should therefore be alert for any indication of distress.

When both sides have been seen to be adequately filled with gas the catheter or needle is removed. The patient is warned to be ready to stand and the table is tilted to bring the patient into a near-erect position. Coned oblique films of the supra-renals are taken, using the under-couch tube. The table is then returned to the horizontal and the patient lies supine.

Before further films are taken, a medium dose of contrast (e.g. 50 ml. Conray 420) is injected and the patient sits upright for about 5 minutes. This is intended to produce maximal separation of the kidneys and adrenals A supine film of the renal areas is then exposed and should be viewed before taking further films. If there is sufficient gas in the region under examination, and if the adrenals are well shown a single film may be adequate. Even then it is advisable to take a prone film as well as a supine so that shadows caused by fluid in the duodenum or gastric fundus will alter.

When there is enough gas but poor visualisation of the organs under investigation, tomography is required and this applies to the majority of cases. A.P. cuts (7–10 cm. for a thin patient, 11–14 cm. for a fat patient) will commonly show adrenals or para-aortic glands. Commonly, especially on the left, an adrenal lying anterior to the upper pole of the kidney does not show satisfactorily. Such cases are best demonstrated by high oblique or lateral tomography (50–60° oblique with the side of greatest interest lowermost). To show the pancreas, lateral tomography lying on the right side is the most useful. In any of these examples it may be of value to 'label' neighbouring organs, e.g. by giving gastrografin or barium or by urography. If suitable tomographic equipment is available tomography is best performed in the erect or near-erect position since the kidneys descend while the adrenals are relatively fixed.

Retroperitoneal insufflation.—If it is only necessary to show one side, e.g. before renal biopsy or percutaneous pyelography, the gas can be injected by the trans-lumbar route. The patient lies prone and the skin is cleaned and the back towelled. The needle is introduced, using local anaesthesia, just above the iliac crest at its highest point. As the needle is advanced deeper, injecting the local anaesthetic progressively, it will enter the retroperitoneal space and the resistance to injection will suddenly drop. After introducing further saline to expand the retroperitoneal space around the needle, nitrous oxide—400–500 ml.—is injected to outline the kidney and suprarenal. After screening to check that this is passing satisfactorily around the kidney, air or oxygen may then be injected if required.

Difficulties and complications

1. *Penetration of the rectum.*—This is most liable to occur if the patient is very fat—since this makes correct direction of the needle more difficult—or very thin, because the presacral space is then narrow. It is also more likely to happen if the needle is advanced without previous injection of dilute lignocaine or saline to expand the presacral space. If penetration is known to have occurred the examination

should be deferred both because of the risk of infection and because of the liability of the gas to leak.

2. *Discomfort in the chest and neck.*—This, which is due to surgical emphysema, occurs following injection of oxygen or air. It is less distressing if the patient is warned of it beforehand. It may appear from $\frac{1}{2}$ to 6 hours or more following the injection.

3. *Gas-embolism.*—This usually occurs soon after the beginning of the instillation of gas and results from inadvertent puncture of a vein by the needle. It may also occur later in the examination if the dissecting gas tears adhesions in the retroperitoneal space.

Many of the points of technique mentioned are designed to diminish the risk of gas embolism. If a vein is punctured during introduction of the needle, the examination should be postponed until later, unless CO_2 or N_2O *alone* are used when this rule does not apply.

Treatment of gas-embolism is discussed in Chapter 3.

4. *Local sepsis.*—Judging from the rarity of reports this is surprisingly infrequent. As the presacral area is relatively avascular it is important to avoid infection since anaerobic organisms might flourish.

5. *Difficulty in outlining the adrenals.*—It is often impossible to achieve satisfactory outlining of both adrenals though the use of tomography, particularly oblique tomography, is of considerable assistance. The glands are surrounded by fat and gas does not invariably 'dissect' cleanly. Patients with Cushing's syndrome may provide especial difficulty in this respect; their retroperitoneal space contains an excess of fatty tissue in which the gas can become unevenly loculated.

6. *Phaeochromocytoma.*—This examination can cause a worsening of symptoms in patients with phaeochromocytoma; suspected cases should be treated with phenoxybenzamine (and propanol if necessary) for two or three days beforehand until the blood pressure has been stabilised. During the examination they must be carefully observed and their blood pressure recorded at intervals during the examination. Treatment with phentolamine may become necessary and it is an essential precaution to insert a catheter into a vein before beginning the study. This allows immediate injection of any necessary drugs without the delay inherent in finding a vein in a collapsed patient (see also Chapter 3).

References

SOWERBUTTS, J. G. (1959). *J. Fac. Radiol.*, **10**, 201.

APPENDIX

Ward Instructions

The examination is at .on. . . . .

1. The patient should have nothing to drink for 12 hours before this and nothing to eat for 6 hours.

2. Give 2 Dulcolax tablets at 8 p. m. the night before and a Clysodrast or trickle enema one hour before the examination.

3. Shave the natal cleft and peri-anal area.

4. The patient should empty the bladder immediately before coming down.

5. Give .... as premedication at .... a.m./p.m.

Trolley setting

As for pneumoperitoneum using an arteriogram needle, needle-cannula (M.D.) or needle-catheter (Longdwel B.D.) 6719 together with the following additions:—

Upper shelf

1 × 20 ml. syringe (for retroperitoneal saline injection).
Arteriographic connection.

Lower shelf

Plextrocan—needle-cannula (or equivalent) with 2 ml. syringe—for venous cannulation in suspected phaeochromocytoma.
5 × 10 ml. saline ampoules.
Nobecutane spray.

CHAPTER 22

PNEUMOMEDIASTINOGRAPHY
Louis Kreel, M.D., M.R.C.P., F.F.R.

Gas contrast visualisation of the mediastinum is particularly applicable to the anterior compartment for the demonstration of the size, shape and position of the thymus, but can also be used for the localisation of an intrathoracic parathyroid tumour and lymph nodes and has also been used for outlining the outer wall of the oesophagus.

Indications

1. In myasthenia gravis where no tumour is visible by conventional radiography or fluoroscopy, especially in cases with positive muscle antibodies.
2. In primary or 'tertiary' hyperparathyroidism where a tumour cannot be located in the neck.
3. It may be used to demonstrate mediastinal lymph nodes.
4. In oesophageal carcinoma to determine the extent of the lesion and to assess operability.

Contra-indications

This examination should not be carried out in patients with a diminished respiratory reserve, particularly if there is any diminution of the expulsive force on coughing. Even a minor attack of coryza in a myasthenic is a contra-indication to the procedure.

Special Points in Booking

The patient should always remain in hospital for 48 hours following the procedure and in the case of myasthenia gravis must be in-patients in a special unit and stabilised on anti-cholinergics. The correct frontal and lateral tomographic exposures are established the previous day. The bladder and bowels should be empty before the patient comes to the X-ray department. The patient's anti-cholinergic drugs must accompany the patient to be given whenever they fall due even if this is during the procedure. The case notes and a consent form should also be available.

Sedation

In apprehensive patients 50 mg. of amytal may be given, but respiratory depressants such as morphine or pethidine must be avoided.

Summary of the Procedure

1. Careful explanation to the patient of the procedure and its effects.
2. Locate sterno-manubrial joint by radiography using a lead marker.
3. Local anaesthetic to skin and subcutaneous tissue.
4. Introduce needle with stilet through manubrio-sternal joint during forced expiration so that its tip is just beyond the retrosternal fascia and check position radiographically.
5. Aspirate to check that no blood returns and then inject 5 ml. of normal saline.
6. Slowly inject 100 ml. of air and take a test film—lateral of anterior mediastinum.
7. Slowly inject further 200 ml. of air and repeat test film.
8. Up to 500 ml. of air may be required but check visibility of thymus with needle in position.
9. Remove needle and take spot films in supine oblique and lateral positions.
10. Frontal and lateral tomography of anterior mediastinum is then carried out.

Technique

1. The examination must be carried out in the radiology department with fluoroscopic control preferably with image intensification and television monitoring. As the patient comes to the department make sure that anti-cholinergic tablets or ampoules have come with the patient who must be free of upper respiratory tract infection and have a good expulsive cough. The patient should then be given a full explanation of the procedure and be warned that a feeling of tightness in the chest will occur and that the following day there will be some puffiness in the neck accompanied by a crackly sensation on palpation.
2. The patient is then placed supine on the radiographic table and a small lead strip is placed over the manubrio-sternal joint and its position marked on the skin. A lateral film of the anterior mediastinum is taken. This is viewed to ascertain the exact position of the joint.
3. After preparation with chlorhexidine/spirit, the overlying skin and subcutaneous tissue is infiltrated with 1% lignocaine down to the joint.
4. A 5 cm. long 20 gauge needle with the stilet in position is then inserted in the midline, in the line of the joint in full expiration. There should be a slight resistance as the needle traverses cartilage. It is inserted until the tip of the needle is just beyond the retrosternal fascia (Figure 26) which is approximately 2·5 cm. from the skin surface. The position of the needle is checked by a lateral film of the anterior mediastinum.

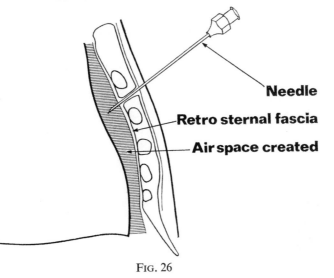

Fig. 26

5. The stilet is removed and a check aspiration ensures that a small
   vessel has not been entered. 5 ml. of normal saline is injected to
   displace surrounding vessels.
6. 100 ml. of air is injected slowly over a period of 5–10 minutes so that
   minimum discomfort is produced; then a lateral check film is taken.
7. A further 200 ml. of air is injected very slowly over about 10–15
   minutes, again ensuring that not too much discomfort is produced.
   The stilet is then replaced and the patient is screened in the lateral
   and oblique positions to see how well the thymus has been outlined.
8. Up to a further 200 ml. of air may be injected slowly, with a further
   check fluoroscopy at 400 ml.
9. The needle is removed and the skin sealed with a collodion spray.
10. Spot films of the thymus are taken in the oblique and the right and
    left lateral positions, carefully marking each film as to side. The size,
    and shape of the thymus should be clearly seen on these films.
11. Lateral tomography of the anterior mediastinum is carried out. The
    arms are best placed above the head. The tube is centred over the
    anterior mediastinum at the level of the manubrio-sternal angle and
    tomographic cuts taken at the midline and at 1 cm. intervals up to
    3 cm. away from the midline on both sides. A multi-section box is
    particularly useful.
12. Prone frontal tomography of the anterior mediastinum is carried out
    centring on the spinous process of D4. The cuts need not extend
    more posteriorly than 9 cm. behind the sternum.

Special problems

The above technique is particularly directed towards demonstrating the
thymus, but applies equally when searching for an anterior mediastinal

parathyroid adenoma or outlining anterior mediastinal lymph nodes. However, for showing hilar lymph nodes or the outer wall of the middle and lower oesophagus the patient must lie in a prone position for 1–2 hours to allow the air to pass into the posterior mediastinum. In the case of the oesophagus a mouthful of barium should be swallowed prior to the appropriate tomographic cuts. For examination of the upper oesophagus, thyroid and cervical parathyroid glands the appropriate tomography should be done 4–6 hours after the instillation of the air. During this time the patient should be in the erect position.

Complications

1. Pneumothorax. A small pneumothorax may be produced, but this is of no consequence.
2. Any pulmonary collapse in a myasthenic patient following this procedure must be viewed with gravest concern for if not immediately relieved by physiotherapy a tracheostomy is indicated. Unless this is done a myasthenic crisis may ensue. If this procedure leads to pulmonary collapse a tracheostomy will be required for the thymectomy in any case.
3. The other complications are those associated with any gas contrast procedure and due either to the local anaesthetic or gas embolism and have been dealt with in Chapter 3.

References and Further Reading

CONDORELLI, L. 1936. *Minerva Med.*, **1**, 81.
KREEL, L. and JAMES, V. 1965. *Radiography*, **31**, 133.
KREEL, L. 1968. *Proc. R. Soc. Med.*, **61**, 754.

Trolley setting

Top shelf

Sterile tray with lid containing
    2 dressing towels
      gauze swabs 4″ × 4″.
    2 gallipots.
    1 three-way tap, Luer fitting.
    1 Howard Jones needle 6·5 cms,
      short bevel 25° with capped stilet, Luer.
    1 Howard Jones needle 7 cms.,
      short bevel 25° with capped stilet, Luer.
    1 50 ml. syringe, Luer fitting.

Bottom shelf

Chlorhexidine 0·5 %.
Methylated Spirit.
Procaine Hydrochloride B.P. 2 % containing Chlorocresol.
Octaflex plastic dressing.
Disposable syringes 1 × 5 ml. and 1 × 10 ml.
Needles, Scimitar 20 (skin needle)
      Scimitar 1.
Masks.
Doctor's gown and hand towel in sterile pack.

CHAPTER 23

Part I

CONTRAST ARTHROGRAPHY OF THE KNEE
(J. L. Boldero, B.M., B.Ch., F.F.R., and
F. W. Wright, B.M., B.Ch., M.R.C.P., F.F.R.)

THIS is a useful method for demonstrating tears of the menisci, thinned and damaged cruciate ligaments, and defects in the articular condyles due to osteochrondritis dissecans.

Contra-indications
1. Allergy to contrast media.
2. Local sepsis around the knee.
3. Active arthritis.

Special points in booking.—Patients should not walk or cycle undue distances immediately following the arthrography: transport home may therefore need to be arranged in some cases.

Ward or Department preparation.—The antero-lateral side of the knee is shaved.

Department preliminaries
1. A.P. and lateral films of the knee joint are exposed, or, if they have already been taken, an A.P. is obtained for exposure purposes.
2. While these are being developed the knee is examined and a history is taken from the patient. He is warned about possible sequelae such as swelling or pain, and told to report back to the department or his general practitioner if these symptoms are severe. (This occurs in less than 5% of patients).

Procedure
A. **Contrast.**—4–10 ml. Hypaque 45%, Urografin 60% or Conray 280.
B. **Technique.**—The extended knee is cleaned with a suitable antiseptic, e.g. Hibitane in spirit. Strict asepsis is required and gloves, masks and towels are used.

The puncture is usually made on the lateral side of the joint $\frac{1}{2}''$ lateral to the middle of the lateral border of the patella. The skin and subcutaneous tissues down to the joint are infiltrated with local anaesthetic, e.g. 0·5% Xylocaine with 1/10,000 adrenaline in a 10 ml. syringe; 4 ml. are usually sufficient. 10 ml. of contrast are drawn up into a 10 ml. syringe.

The puncture of the joint is made using the syringe containing the local anaesthetic to which is attached a hypodermic size 0 needle with a short bevel. The patient is instructed to relax his knee so that the patella can easily be dislocated laterally. The needle is inserted to pass behind the lateral border of the patella. After it has penetrated the skin, the patella is dislocated laterally with one hand so as to open up the space behind the patella, whilst the needle is advanced into the joint. It may be necessary to inject a further small amount of local anaesthetic when the joint capsule is reached. Sometimes this is felt with a slight 'give', but usually the indication that the needle is in the joint is that resistance to the injection of the local anaesthetic ceases. If injection is impossible this usually means that its tip is embedded in cartilage; the needle is then withdrawn a few millimetres and re-inserted. When the joint has been entered satisfactorily the syringe is then changed and 4–5 ml. of contrast is injected. The puncture wound is sealed with collodion. (If the joint contains an effusion as much as possible should be aspirated before the contrast is injected to avoid dilution. A wide-bore serum needle is sometimes required as the fluid may be very viscous. To help empty the joint, compression of the suprapatellar pouch and over the patella is employed.)

The knee is put through as full a range of movement as possible several times (flexion, extension and rotation) to distribute the contrast. The joint margins are determined by palpation and the skin marked anteriorly, medially and laterally to facilitate radiographic centring. A crêpe bandage is then applied above the patella to compress the suprapatellar pouch. This is left *in situ* until all the films have been obtained.

Radiography is best carried out on a Lysholm skull table, as it is essential that the tibial plateau should be vertical to the film so that the shadows of the menisci will be projected clear of bone. With this apparatus the height of the table is easily adjusted, and the pointer and/or light collimator assists centring. The field is restricted to a 6″ circle and 6″ × 8″ films are used in the Bucky tray.

Eight films are taken, centring to the middle of the knee-joint.

> One A.P.
>
> Six obliques turning the knee 10°, 25° and 45° to each side.
>
> One lateral with 70° to 80° of flexion of the knee, i.e. with an angle at the knee of 100° or 110°. (This needs about 5 kV increase in exposure.)

The foot should be immobilised in each position, using sandbags.

Extra lateral films may be taken in a case of suspected cruciate injury, each ligament being stretched in turn. A 'sky line' view may also be of value to show the posterior surface of the patella.

Once the contrast has been injected the examination must be completed in as short a time as possible as it is rapidly absorbed from the joint, and films taken more than 10 minutes after the injection are of little diagnostic value.

References

FISCHER, F. K., SCHINZ, et al. (1952), *Roentgen Diagnostics Am. Edn.* Vol. ii, Part 2, p. 1247.

LINDBLOM, K. (1948). *Acta Radiol. Suppl.*, 74.

LINDBLOM, K. (1952). *J. Fac. Radiol.*, **3**, 151.

OLSON, R. W. (1967). *Am. J. Roentg.*, **101**, 897.

Note.—Double contrast or gas arthrography may also be used.

Double contrast

ZAKRISSON, U. (1960). *Acta Radiol.*, **53**, 442.

Gas

BONNIN, J. G., BOLDERO, J. L. (1947). *Surg. Gynec. and Obst.*, **85**, 64.

APPENDIX

Trolley setting

Sterile—upper shelf

Gloves and powder.
3 Towels.
2 × 10 ml. syringes.
syringe.
2 No. 16 needles.
2 No. 1 needles.
2 No. 0 short bevels.
1 Serum needle for aspiration of contrast from ampoule.
1 gallipot for antiseptic.
Forceps to hold swabs for cleansing the skin.
Small forceps for holding needles, etc.

Lower shelf

Local anaesthetic.
Contrast.
Collodion.
Antiseptic.
Masks.

Part II

ARTHROGRAPHY OF THE SHOULDER

(*Lipmann Kessel, M.B.E.(Mil.), M.C., F.R.C.S.*)

CONTRAST-MEDIUM arthrography of the shoulder joint may be used in the diagnosis of injury to the musculo-tendinous cuff: the so-called supraspinatus tendon rupture. It may also be used to elucidate the cause of rare disorders such as chondromatosis, etc. Recently the technique has been extensively used in experimental work to illustrate the precise anatomy of various conditions such as recurring dislocation of the shoulder. With normal care it is an entirely safe procedure without complication. In order to interpret the results of the investigation correctly, an exact knowledge of the appearance of a normal shoulder arthrograph is necessary; any errors of technique, however slight, will invalidate the result and lead to false interpretation.

Contra-indications.—There are no contra-indications, except infection of the shoulder joint or in its region.

Out-patient Preparation.—Usually none is necessary, but in the case of a nervous patient, promethazine, 25 mg. intramuscularly, 45 minutes before the procedure will be found adequate.

Summary of Procedure

1. Plain films of the shoulder joint.
2. Infiltration with local anaesthetic.
3. The introduction of the needle for arthrography.
4. Rotation of the arm to confirm the position of the needle.
5. Introduction of a small amount of contrast medium, with screening or a film to confirm that contrast is entering the joint space.
6. Introduction of the full volume of contrast medium.
7. Removal of the needle and repeat standard views of the shoulder in external and internal rotation, together with an axial view.

Equipment.—The only special equipment required is a needle of fine gauge with a short bevel. The use of a pediatric lumbar puncture needle has been found to be the most suitable. The reason for the choice of this needle is discussed below. A separate syringe and needle are used for local anaesthesia.

Contrast Medium.—As contrast medium we have used 35% diodone, which remains in the joint for ten to fifteen minutes, an adequate period for unhurried radiographs to be taken. 25% or 45% Hypaque are also satisfactory.

Department Preliminaries.—The patient is placed supine on the X-ray table, and plain films of the shoulder are taken on the Bucky, the first in internal rotation and the second in external rotation and in each case a good joint space view is taken at right-angles to the plane of the scapulo-humeral joint. To achieve this the patient (not the arm) must be rotated about 50–60°. The ray should also be angled about 20° towards the feet so as to demonstrate the space between the acromion and the humeral head. The third film taken is an axial view of the shoulder. If one is not familiar with the technique it is useful during the first few investigations to have a screen available so that the position of the needle may be checked by fluoroscopy before the contrast medium is introduced. Once familiarity with the technique has been obtained, such screening is no longer necessary.

The patient then lies supine with the arm to his side and the hand palm upwards.

Technique

The skin of the shoulder region is prepared in the usual way. A small wheal of local anaesthetic is made at a point 2 cm. anterior to the acromio-clavicular joint. The local anaesthetic is then introduced directly on to the head of the humerus; the direction of the needle is down the line of the patient's arm and backwards towards the table at an angle of 45° to the vertical. At this point the head of the humerus lies only some 2 to 3 cm. deep to the skin. Local anaesthesia having been induced, the lumbar puncture needle is introduced without attached syringe, along the same line and *the point of the needle is embedded firmly in to the articular cartilage of the head of the humerus* (Fig. 27). This is the key to a successful examination.

The embedding of the point of the needle into the articular cartilage is confirmed by asking the patient to turn his hand inwards and outwards through a few degrees, when it will be seen that the needle is rocked to and fro through a wide arc because the point is embedded in the head of the

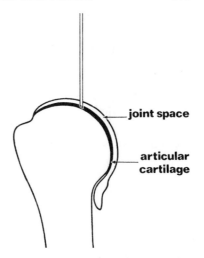

FIG. 27.—Shoulder arthrography. The needle is embedded in the humeral articular cartilage before withdrawal into the joint space. Note that the space shown is a diagrammatic representation; the true joint space is potential only.

humerus. Once the embedding of the point of the needle in the articular cartilage has been checked, the stilet is withdrawn and the syringe containing 8–10 ml. of contrast medium is attached to the needle, which is then gradually withdrawn while pressure is maintained on the plunger of the syringe. It is very important to remember that at the site of the injection there is no more than a capillary joint space. Almost without exception failures are due to the use of a needle which has too long a bevel; it then becomes only too easy to inject contrast medium into the joint, the subdeltoid bursa and the related soft tissue planes at one and the same time, thus invalidating the investigation. Because the joint space is only of capillary size the contrast fluid cannot be injected into the joint *on the way in* because there is no way of knowing when one has reached the joint space itself. The only way to do it, therefore, is to put the needle beyond the joint space into the articular cartilage and to inject *on the way out*, by gently withdrawing the needle while maintaining pressure on the plunger of the syringe. In this way a pure intra-articular injection can regularly be obtained.

At first only 1 ml. of contrast is injected. The shoulder is then screened or a film taken in neutral rotation with the needle *in situ* to check that the joint space has been entered and that there has been no leak back at the site of needle puncture. If a control film of this sort is used it is advisable to overexpose it so as to shorten the developing time.

Provided the position of the needle is satisfactory, the remainder of the contrast is injected. The definitive radiographs are taken after this when the needle has been removed and all the required information may be obtained from the three views described above.

Screening during the injection is desirable at least until one becomes familiar with the technique. If the injection is extra-articular, the leak will be seen during the injection and the procedure can then be stopped and repeated on another occasion. Image amplification is, of course, an added advantage.

Complications and Difficulties.—These should not arise with a technically efficient, entirely intra-articular injection. If a false injection has been made, that is if a simultaneous injection into joint space and extra-articular tissues has been carried out, the patient will complain of pain. The injection should be stopped forthwith, because the complaint of pain is an indication of a false injection. Usually no pain is experienced by the patient, although some mild ache may be felt extending down the arm.

References and further reading
ELLIS, V. H. (1938). *Proc. R. Soc. Med.*, **31**, 451.
ELLIS, V. H. (1952). *J. Bone Jt Surg.*, **34B.**, 513.
KESSEL, LIPMANN (1950). *Proc. R. Soc. Med.*, **43**, 418.
LINDBLOM, K., and PALMER, I. (1939). *Nordisk Med.*, **1**, 532.
LINDBLOM, K. *Acta chir. scand.*, **82**, 133.
NELSON, D. H. (1952). *Br. J. Radiol.*, **25**, 134–140.

APPENDIX

Trolley setting

Sterile trolley

    Gauze swabs. Dressing towel.
    2 × 10 ml. syringes.
    Needles, hypodermic No. 20 and No. 1.
    2 gallipots.
    1 pediatric lumbar puncture needle, short bevel.
    Luer-Record adaptor (if required).

Lower shelf

    Bottles of iodine, spirit.
    Bottles of lignocaine 1 %.
    Ampoules 35 % diodone, or
      25 % or 45 % Hypaque.

CHAPTER 24

SINOGRAPHY AND FISTULOGRAPHY

THESE examinations can be of great assistance to the surgeon but they should never be embarked upon without a complete understanding of the clinical aspects of the problem. It is vital to know what the clinician suspects or wishes to discover; equally a knowledge of past history or previous operations may suggest a different approach or different views. The challenge presented by a difficult sinogram may be considerable, requiring perseverance and a very careful assessment of the problem under investigation.

Contra-indications

1. Heavy infection of the sinus in the presence of pyrexia may contra-indicate examination.
2. Poor general condition of the patient occasionally precludes the investigation.

Ward preparation and premedication.—Pain relieving drugs such as pethedine may be needed, otherwise no preparation is used. Whenever possible the radiologist should see the patient before the investigation to examine him clinically and to decide the best method of introducing the contrast medium.

Summary of the Procedure

1. Screen the region to be examined or take plain films.
2. Clean the sinus mouth and choose an appropriate introducer to fit it.
3. 'Label' the sinus mouth(s) and any other surface markings as desired.
4. Position the patient for the first film (sinus uppermost). Attach introducer to a syringe of contrast medium, fit it into the sinus mouth and seal with moderate firmness.
5. Inject until (*a*) Contrast medium spills back.
    or (*b*) the patient complains of pain
    or (*c*) the resistance to injection rises
    or (*d*) 40–60 ml. have been injected.
6. Wipe away any contrast that has been spilt and take films; if contrast continues to spill out inject more before the second film is taken.
7. Remove the introducer; mop any further spill of contrast; await films.

Contrast media

The range of opaque media is very wide but it is only necessary to consider a few different types and the problems in which they are most suitable.

Iodised oils.—The use of these media has diminished with the advent of the more opaque water-soluble media. Neo-hydriol is a satisfactory oily medium. It may be used to fill cavities where it is desired to have a relatively permanent residue. This sometimes applies to empyema cavities but if there is a suspicion of a broncho-pleural fistula it should not be used as it leaves a permanent residue in the alveoli.

Bronchographic media.—These media are suitable for outlining empyema cavities and broncho-pleural fistulae. Dionosil oily or Hytrast are current examples of this type of medium.

Water-soluble media.—Of these we have found Urografin 76% Conray 420 and Hypaque 85% the most generally useful media. They are harmless to the tissues and are absorbed quickly, leaving no inert residue. They are of suitable viscosity; they mix with body fluids without globulating and are reasonably opaque. They are the media which we normally use for sinography and fistulography but should not be employed in empyema cavities if there is any possibility of a broncho-pleural fistula since they cause severe distress and coughing if they enter the bronchial tree.

Umbradil viscous U.—for urethrography—we have found of value for short fistulae which tend to empty quickly, e.g. peri-anal fistulae.

Equipment

To inject a sinus with contrast some form of introducer is required. Four types will be described:—

1. **The fine cannula.**—Where the sinus mouth is very small, as may happen in the face and neck, a small cannula—such as a lacrimal cannula or that used for sialography—is the best introducer. It may be attached to the syringe directly or via an arteriographic connection (p. 194).

2. **A tube.**—This may vary from fine polythene or nylon (e.g. epidural cannula) to wide rubber 5 mm. or more in diameter. Each requires some form of adaptor to connect it to the syringe. The largest tube which will fit the sinus mouth is chosen and is inserted as far as possible. To produce a more effective seal the patient may press down with his finger over the tube and sinus mouth; or this can be done with a swab by an assistant or the operator who will remove his fingers just before the exposure. When a track is tortuous or the tissues lax it is possible to advance a tube so far that its end is occluded or the track kinked and the sinus then fails to fill. When this is suspected the tube is withdrawn a little. In lax tissues it may be necessary to draw the tissues outward between finger and thumb so as to straighten the track (Gage and Williams, 1943).

3. **Nozzle type introducer.**—For many sinus mouths a rubber urethral nozzle is convenient especially if mounted in a suction cup and connected to a length of tubing to enable the syringe to be kept outside the area of irradiation. The latter type, the Pridie injector, is useful but cannot always be employed particularly when the sinus mouth is irregular due to distortion from scarring.

4. **The balloon catheter.**—Alternatively for large orifices a Foley catheter may be used. The balloon is blown up firmly and the catheter tip is then introduced into the sinus mouth. The patient or an assistant presses down on the balloon which seals off the sinus very effectively (Fig. 28). (This method—using the larger self-retaining enema catheter—is also useful in giving barium enemas to patients who have a colostomy.)

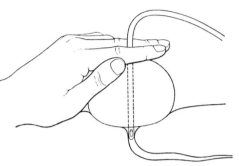

FIG. 28.—Sinography. Use of a balloon catheter for injection of sinuses. Firm pressure by the patient or operator produces an effective seal (also of value in colostomy enemas).

Technique of Obtaining Films

The choice is between screening during introduction of contrast and 'blind' introduction with overcouch films.

The advantages of screening with image intensification are:—

1. Accurate positioning of films, remembering that sinuses may run in unexpected directions.
2. If there is a fistula, e.g. into bowel, it will be quickly seen and time will not be wasted injecting an unnecessary amount of contrast through it. Moreover, when there are large cavities their size will be appreciated in good time.
3. The patient is more easily brought into the erect position.

The disadvantages are:—

1. It is more difficult to maintain any kind of asepsis at the sinus mouth.
2. It is more difficult to prevent leakage at the sinus mouth.
3. Irradiation of patient, operator and assistant is increased.
4. The patient must be turned for views at a right angle (see Special problems).

When a sinus is thought to arise in bone, overcouch films have the advantage of sharper definition; for examination of fistulous tracks screening has distinct advantages: for most sinuses either method is adequate.

G

And with image intensification it is possible to achieve a combination of both methods, injecting contrast with room lights on, screening to confirm adequate filling and then taking films with the overcouch tube.

Technique

1. All removable dressings and any safety pins attached to them are removed. Safety pins through drainage tubes are, of course, left in position.

2. The area to be examined is screened or plain films are taken, the latter usually at right-angles. Their purpose is to show abnormalities within the patient, e.g. opaque foreign bodies or lesions in bone. They also provide a check on the correctness of the exposure. When a sinus is thought to be related to bone it is important to secure good quality films of the region and these are best taken some time before the examination so that they may be studied at leisure, (Gage and Williams, 1943.).

3. The patient is positioned appropriately on the couch. This usually means with the sinus mouth uppermost since otherwise contrast tends to drain from the sinus and also the sinus becomes less accessible.

4. When injecting a space which already contains a drainage tube (e.g. an empyema) a Jacques catheter is passed into the drain. The patient is positioned with the drain uppermost. Contrast medium may then be injected until it spills freely, often bringing with it some pus. This contrast is then mopped away and a pair of artery forceps is applied to the drain and the tube; the forceps are slowly tightened during the injection of a further contrast. Contrast is injected until it refluxes around the *outside* of the tube, so ensuring that the sinus or cavity is completely filled. It may be necessary to pull the drain slightly outward to ensure that its tip is not impacted against the side of the sinus in which it lies, so preventing adequate filling. An alternative method is to introduce a small balloon catheter, inject contrast to fill the tube and then inflate the balloon so that it grips the tube. Contrast is again injected until it spills out around the tube.

   When the cavity is large it will be necessary to roll and tilt the patient so as to coat the walls with contrast medium since it will not be possible to fill the space with the usual maximum volume of 40–60 ml. With this method it is important to be certain that the drainage tube does not lie with its end against the wall of the empyema space since a false impression of the size of the cavity will be gained. If necessary the tube should be partly withdrawn.

5. For orifices with no drain in them an appropriate introducer is employed. First the sinus mouth is carefully cleaned with Savlon or similar fluid. Some sinuses have very small orifices which may be covered by crusts or scabs. These will be wiped away gently. If the sinus mouth is still difficult to find, massage of the surrounding tissues

in its direction will often produce a small efflux of pus. Although there are certain dangers in probing a sinus—trauma to the walls and creation of false passages—gentle probing can be very useful in determining the direction in which to introduce a cannula or tube. The introducer which has been chosen is fitted into the sinus mouth to make sure it is satisfactory.

6. Any lead letters to be used in marking (see below) are cleaned with spirit and stuck (using plaster) on to the skin next to the sinus mouths or in other chosen positions.

7. A syringeful—normally 20 ml.—of contrast medium is drawn up; it is connected to the introducer and bubbles are expelled.

8. When the overcouch method is used a film is put in the Bucky tray and the tube centred.

9. The introducer is put into the sinus mouth, sealed off and injection commenced. When sufficient is judged to have been injected (see below) a film is taken. If there is a spill of contrast it must be wiped away before the exposure. During the actual exposure, however, the operator should take his hands away so that a small amount of spill may occur. The patient is then turned preferably through a right angle, a little more is injected and a second film taken. Any oblique films which are thought advisable may be taken at this stage although more often they are taken after the initial films have been viewed. The introducer is removed, excess contrast mopped away and the films are awaited.

When the films are viewed they must be carefully considered as to whether:—

(*a*) they are radiographically satisfactory;
(*b*) the sinus or fistula has been shown in its entirety;
(*c*) oblique, erect views or head down prone or supine views might help further or whether more contrast or a different contrast should be injected.

The patient is sent back to the ward only when the operator is satisfied that the problem with which he was presented has been resolved.

Control of injection.—This presents a certain difficulty. To show all the ramifications of a sinus or fistula it is usually necessary to raise the pressure within the sinus at some stage, which involves sealing off the sinus mouth. Yet too high a pressure may force infected material into the surrounding tissues or blood stream or may even create fresh extensions of the sinus. One way of controlling the pressure is by using a funnel or using the barrel of the syringe in the same way as a funnel. For most sinuses, however, it is sufficient to inject gently by hand with a moderately firm seal at the sinus mouth: injection is then stopped when resistance rises, when the patient complains of pain or when contrast leaks out of the sinus mouth. To maintain injection during exposure and so to provide a continuous column of contrast—as recommended by Gage and Williams (1943)—is undoubtedly the ideal method but may involve excessive irradiation to the operator.

What films to take.—An AP (or PA) and lateral are standard, but intelligently chosen oblique films are often valuable. Less often stereoscopy and tomography may be helpful, especially with sinuses related to bone. With large cavities an erect pair of films or a film with horizontal ray may be useful.

Chest films are taken in inspiration, abdominal films in expiration. A helpful pair of films for empyemas are a PA prone or AP supine with the space filled, followed by an erect lateral after unclipping the drain so that the natural level in the empyema is shown.

When the examination is conducted 'blind' it is usually advisable to use the largest possible films since sinuses may run for considerable distances and to 'cut off' the end of a sinus is most exasperating.

Marking.—Every sinus mouth should be marked with a lead letter. The letter O with the tube through its centre will mark the main injection site. Where multiple sinuses are present each is numbered and the system of marking is described in the report, if necessary with a diagram.

The other use of markings is to show the relationship of sinus ramifications to surface markings on the skin. This is of particular importance in the abdomen; for example with a sinus laterally in the abdomen and known to be running forward, a marker was put on the umbilicus and another on a point (marked indelibly) half-way between xiphoid and umbilicus; this gave assistance in localising the sinus anteriorly. When screening is used, of course, such markers may be placed precisely over the sinus under screen control but the skin must still be marked indelibly before the patient is returned to the ward.

Occasionally more complex methods of markings are used, e.g. an I.V.U. to show the kidney, Gastrografin swallow to show the stomach. In perianal problems the anus may be marked by inserting into the rectum a rectal tube coated with Microtrast (Damancy).

Special problems.—1. With very small fistulae or with large sinuses, control of injection by the method described may be less satisfactory. Injection of narrow fistulae may give rise to considerable pain without the fistula being completely filled. In such cases screening is an advantage; otherwise it may be necessary to take a film to show where the contrast has reached and to help decide whether more is needed. Difficulties can also occur with large sinuses or with fistulae into the gut if screening is not employed. There is then no rise in pressure or backflow to give guidance as to when injection should be stopped and films are best taken when about 40 to 60 ml. have been injected, judging from these films whether more contrast is required. In fact, if the patient is suitably postured to coat the walls of a cavity, it is seldom that more than 100 ml. is necessary. Very large volumes indeed may, however, be needed when there are multiple large branching sinuses as opposed to one large cavity.

2. Occasionally a difficulty arises in a patient with multiple superficial sinuses which are suspected of having a deep connection. Contrast injected into one superficial orifice is likely to spill rapidly from one or more of the other openings. In this situation the possible approaches are:—

(a) If the deep connection can be found on inspection or by probing, a tube may be introduced directly into it—by-passing the superficial tracks.

(b) Should no deep connection be apparent, injection is made into the main orifice while an assistant occludes the other sinus mouths by pressing on them with the fingers (gloves are desirable). Usually it is rapidly evident whether the sinus is only superficial—in which case the pressure will rise uncontrollably—or whether the contrast is passing deeply.

3. One further point may be mentioned. In most patients a lateral view is taken with the subject lying on his side. Where the sinus is wide mouthed or there are multiple openings the patient must be kept with the orifice(s) uppermost or the contrast will drain away. Such patients are therefore laid in an appropriate position on a 6″ polyfoam mattress or other thick padding. They lie close to the mattress edge which is furthest from the X-ray tube. A grid and a cassette may then be placed against the patient for a view taken with horizontal ray. This enables films to be taken at right angles without rolling the patient.

Complications.—Spread of infection is the most important hazard and may be avoided by gentleness during injection. It should be remembered that the organisms at a sinus mouth may well include varieties not found within the sinus and careful cleaning is therefore desirable to avoid introducing fresh bacteria. The creation of false passages is another complication which is usually the result of using excessive force in probing or injection.

The use of oily media provides a potential danger of oil-embolism and for this reason it is preferable to restrict the volume employed to 2–3 ml. whenever Neo-hydriol is injected under any pressure.

Reference
GAGE, H. C., and WILLIAMS, E. R. (1943). *Br. J. Radiol.*, **16**, 8.

APPENDIX

Trolley setting

Sterile trolley

3 towels; gauze and wool swabs (plentiful).
1 gallipot with chlorhexide/spirit.
1 gallipot with Savlon.
1 receiver with 1 × 20 ml. syringe and drawing-up cannula.
No. 1 needle.
1 pair sinus forceps.

1 probe.
1 pair artery forceps.
1 spigot.

Lower shelf

Contrast medium—Neo-hydriol
Urografin 76%
Conray 420   } in warm water
Hypaque 85%
Dionosil oily
Umbradil viscous U

Box of lead letters.
Roll of adhesive plaster.
Disposable gloves. Paper hand towel.

Sterile packs

Bardex adaptor
2-way tap-adaptor
Arteriographic connection
Assorted Jacques catheters
    ,,   Foley   ,,
Scissors
Pridie injectors (Down Bros.)
Lacrimal and sialography cannulae.
Epidural catheter (for small sinus openings).

CHAPTER 25

VASCULAR RADIOLOGY—AN INTRODUCTION
(*Excluding cerebral angiography and angiocardiography*)

THIS is now a well-established and flourishing branch of radiology. The range of problems to which it is applied continues to grow and there is a steady flow of improvements in technique and equipment. Since the introduction of percutaneous catheterisation by the Seldinger (1953) technique, however, there has been no major change in fundamental principles and it is chiefly these which we describe in the ensuing accounts. We have set certain arbitrary limits to what we have written since it is not our aim to give a complete account of vascular radiology. What we have attempted is to provide the details which are taken for granted in most works on this subject and so to provide a sound foundation upon which the acquisition of more advanced techniques may be based.

The practice of vascular radiology has certain difficulties and is not always free from hazard but really careful attention to every detail of technique will bring the incidence of significant complications to a very low level. The possible complications are so numerous that we do not propose to describe them all in our accounts; the interested reader should consult Sutton's (1962) book *Arteriography* and Ansell's (1968) paper for a discussion of this subject. Our own techniques as described here are intended to incorporate all possible measures to reduce the risk of each particular procedure. However, it must be emphasised at the outset that the most important factor for safety lies in the operator himself; he must realise when it would be unsafe to go further or when he has reached the limit of his capabilities, and he must be ready to give up at such moments, yielding his pride to the patient's welfare. For this reason we would stress that the accounts which follow are designed to fill out the experience already obtained from watching arteriograms being performed and from performing some arteriograms under supervision. We cannot recommend starting arteriography without such a background.

Training of staff

We must also emphasise the importance of the preliminary training of radiographers. They should be completely familiar with the apparatus used and should practice their various duties repeatedly with numerous 'dummy runs' before a patient is examined arteriographically. It is desirable, when possible, to have nurses who are trained to look after needles, connections, catheters, etc., and to assist at the arteriograms. When the X-ray department does not have its own nurse it is usually possible to arrange for assistance from other departments, e.g. Out-Patients or Theatre. At least two nurses should be trained, to provide cover for holidays and sickness.

Collaboration with anaesthetists—resuscitation equipment

On the occasions when a general anaesthetic is needed the radiologist should ensure that the anaesthetist is familiar with certain problems peculiar to vascular procedures. When the examination is of the abdomen he should be warned not to insufflate the lungs with a face mask during induction—this almost always fills the stomach and later the small bowel with gas. For any procedure requiring arrest of respiration a cuffed endotracheal tube should be used and the need for prolonged respiratory arrest must be stressed. Any anaesthetist undertaking anaesthesia during angiography for the first time should be made aware that contrast injection can cause involuntary movement in a lightly-anaesthetised patient.

Should an emergency arise during a vascular investigation conducted under general anaesthesia the anaesthetist will provide experienced help, but when local analgesia and sedation are used the radiologist and his staff must treat the patient. Accordingly adequate resuscitation equipment is required and as elsewhere, all staff should be familiar with its use.

The Patient, the Ward and Arteriography.—At best an arteriogram is not a pleasant experience and it is, therefore, important to consider the patient's feelings very carefully, applying the principles discussed in Chapter 1 and ensuring that sedation is adequate. In Chapter 1 and in the sections concerned we mention the need to warn patients of the sequelae of examinations—local pain and stiffness, bruising and sometimes swelling. It is also advisable to explain to the patient that the procedure may be varied. Otherwise, for example, a patient due to undergo Seldinger puncture in the groin may be disturbed when axillary puncture is also performed. In all cases written consent should be obtained, as mentioned in Chapter 1. The standard consent form for operation is not ideal for this purpose but provided the accompanying explanation is adequate, it will suffice. In addition, even when the radiologist can visit the patient in the ward to examine him and explain the arteriogram, it is desirable that ward staff—housemen, sisters and nurses—should also be able to answer questions about the procedures. To this end we recommend that representatives of ward staff should visit the X-ray department to see arteriograms being carried out and to be adequately 'briefed'. A further reason for ensuring that housemen and ward sisters are well-informed about vascular procedures is to make sure that they are aware of possible complications after the examination. The aim should be to achieve a matter-of-fact attitude which does not magnify the examination into an ordeal to be dreaded, but equally does not gloss over the fact that when the examination is conducted under local analgesia there may be moments of pain. In this it may be helpful to liken the procedures to a visit to the dentist—something which is rarely enjoyed but which most people are able to endure.

The written instructions which are sent to the ward when the patient returns there are based on the sections headed After-care. There is no doubt that in addition to this the radiologist should visit the patient in the ward 4–6 hours after the procedure and if necessary on the day following arteriography to look for sequelae and provide any reassurance which may be needed.

Apparatus for obtaining multiple films

It is possible to obtain useful, informative films without any special apparatus, and we will discuss briefly some of these methods. But any department which undertakes more than the occasional arteriogram or venogram should acquire some form of film changer. This is not only because of its great convenience but because it is of fundamental value in interpretation to see different phases of arterial or venous filling. For a department which performs arteriography with any frequency, a rapid film changer such as the Schonander (AOT) machine is a necessity. Those departments which only undertake occasional arteriography will usually find one of the simpler changers to be adequate but it should have the following minimum attributes:—

1. It should take at least four films in succession at intervals of not more than $1-1\frac{1}{2}$ seconds.
2. It should be possible to move it, linked to the tube, or better still to move the patient over it, in order to cover different areas of the body in succession.
3. It should be simple in design and light in operation so that radiographers can use it without undue effort.
4. It should use cut film or cassettes in order that the films may be developed and viewed as soon as possible after exposure. It must be stressed that whenever a system using multiple cassettes is employed *it is most important to ensure that the screens match one another*; otherwise inequalities of exposure will result.

Some of the simpler methods which may be employed before a changer has been obtained are discussed in the Appendix.

Preliminary films.—These are required to check the position of film and tube for the areas under examination, and to obtain the correct exposure at each site. This is most important; too dark a film will 'black out' the vessels in the soft tissues and too pale a film will cause the artery to be masked where it lies behind bone. Choosing the correct exposure is to some extent a matter of experience but a useful rule in limb arteriography is that one should be able to make out both the edge of the muscle shadows *and* the trabecular pattern of the bone.

Timing of exposures.—Using the simpler forms of film changer there is relatively little difficulty in timing a film series when all the exposures are to be made in the area where contrast is being injected. It is then only necessary to ensure that the first exposure is made before the injection of contrast is completed. Usually if the two subsequent exposures are made as rapidly as possible and the fourth after an interval of 3–4 seconds, a diagnostically satisfactory series will be obtained. With the rapid film-changers such as the Schonander (AOT), films may be exposed at half-second intervals beginning shortly after the start of the injection and continuing for four seconds; films thereafter are exposed at one-second intervals for up to 4–6 seconds. This sort of cover is usually more than adequate for the majority of vascular problems encountered.

The real problems in timing arise when attempting to show an entire limb by serial radiographs or when the examination centres on a part of the limb some distance from the point of injection. With image intensifier fluoroscopy an injection of contrast medium can be watched on the monitor screen and timed as it passes along the arterial tree; or the moment of its arrival at certain key points, e.g. the knee in lumbar aortography, can be timed. In either case the time of exposure of the subsequent film series is adjusted accordingly.

An alternative method,[1] particularly useful in the absence of image intensification, is the injection of a radio-isotope, e.g. Hippuran (sodium orthoiodohippurate containing 50 microcuries of I^{125} in 8 ml. saline) with suitable counters placed distally to time the moment of its arrival and of its greatest intensity.[2]

Calling for exposures.—Usually the operator calls 'take' for each exposure or for the initiation of the series when an automatic changer is employed. When a hand changer is used the radiographer who is making the exposures calls 'pull' immediately after she has exposed. This is also done during a moving series using individual exposures with an automatic changer. This tells the assistants who may be moving the tube (or the patient) or changing the cassettes to act at once. All calls should be loud and clear. It is most important that the various manoeuvres, including moving the tube, cassette changing and exposure alteration should be repeatedly practised until they have been perfected. Any department which purchases an automatic changer should send the radiographer in charge of it to learn the details of its use from an experienced team.

Injection of contrast.—Manual injection of contrast is adequate for all peripheral arteriograms. We prefer to inject direct puncture lumbar aortograms by hand except after translumbar catheterisation. For hand injection the use of a 'guard' on the barrel of the syringe is an important safety precaution with glass syringes. Alternatively metal or plastic

[1] On occasions it may be desirable to establish the position of an arterial catheter or needle when neither X-ray nor isotope control are available. Fluorescein may be injected as 5 ml of a 10% solution (for the aorta) or in 5% concentration for peripheral vessels. Injection is made in total darkness with an ultra-violet light shining on the appropriate limb which should have been prepared by warming. The sudden appearance of yellow fluorescence will indicate the correct siting of the catheter; for example, a catheter passed up from the right femoral artery can be shown to have reached the aortic bifurction when injection causes fluorescence to appear in the left leg.

[2] When this method is used the radio-isotope remaining in the arterial connection at the end of injection is drawn back into the syringe and this is returned to the Physics or Isotope Department for disposal.

syringes may be employed. For catheter injection into the aorta the best results are obtained with a pressure injector, except in children where hand injection usually gives adequate filling. Pressure injection is also desirable for some selective visceral arteriograms, particularly the demonstration of the coeliac axis and superior mesenteric arteries. Most other branches of the aorta can be well shown by hand injection. The pressure injector chiefly employed in general hospitals in this country is the Talley pneumatic injector: it is reliable, efficient and simple to use. Details of its use are given in part IV of Chapter 29.

APPENDIX

Trolley settings for vascular investigations

Methods of preparation for such trolleys will vary in different hospitals with the availability of nursing staff and the methods of sterilisation employed in the hospital. In general there are advantages in preparing a basic pack containing the towels, gallipots and bowls in common use. This can be autoclaved in a paper bag or other suitable container and is then opened out to form the basis of the trolley. Other items, packed individually in paper bags and autoclaved, are then added as required.

The pack should be wrapped twice with water-proof paper (or one layer of water-proof paper in a paper-bag or outer cover). The outer wrapper/bag should be discarded at the time of preparing the trolley. Equipment on the trolley should be kept to a minimum. When preparing the trolley a routine pattern of placing equipment should be established; this prevents any waste of time in searching for a particular item on the trolley. Additional sterile articles likely to be, but not routinely used, should be readily available in packs on the bottom shelf of the trolley.

To prevent confusion with regard to colourless lotions, e.g. local anaesthetic, contrast, etc.:—

1. Mark syringes with different coloured threads and retain individual syringes for a particular use.
2. Discard unwanted lotions/drugs immediately after use.
3. If the practice is to pour contrast into a container on the trolley, use a container of different design, e.g. contrast into a medicine glass, skin preparation into a gallipot.

Basic Pack

Inner wrapper and towel to form base of setting.
6″ bowl with gauze squares (swabs 4″ × 4″ are adequate).
8″ kidney dish with gallipot (for skin preparation lotion), medicine glass (for contrast).
$\frac{1}{2}$ pint jug (for normal saline for rinsing needles etc.)
1 or 2 pint jug (for dextrose or normal saline when hand flushing of catheter is employed).
6 towels 36″ × 36″.
1 towel 60″ × 60″ with 2$\frac{1}{2}$″ square aperture suitably placed for puncture of the artery.
1 piece of water-proof paper 20″ × 40″.[1]
1 7″ sponge holder.
1 Bard Parker handle No. 3 with No. 11 blade.
4 cross action towel clips.
Lengths of black, white and coloured thread (for marking syringes).
Complete unit × 500 ml. (Steriflex) dextrose 5% or dextrose-saline for flushing cannulae, needles or catheters (used with Fenwall pressure bag).

[1] *Water-proof Paper*, 40″ × 40″ sheets. From: Bowater Scott, Bowater House, Knightsbridge, London, S.W.1.

Basic Setting

The "pack" as above, plus:
2 × 20 ml. syringes (for contrast).
2 × 20 ml. syringes (for flushing lotion).
1 × 10 ml. syringe (for local anaesthetic).
1 each No. 1 and No. 17 injection needles.
1 drawing-up cannula.
Sterile autoclave tape.

Articles to be added from separate packs when the details of the procedure are known:
Puncture needle.
Arterial connection.
Syringe for injection of contrast during series of films, e.g. Talley pressure syringe, etc.
Articles necessary for Seldinger catheterisation—see p. 235.
Cover setting with sterile towel on top of which place sterile gowns, gloves and hand towels for the operator and assistant(s). Cover again with a sterile towel, unless the setting is to be used immediately.

The lower shelf, or where possible, a separate trolley close to that containing the 'Basic Setting'.

Skin preparation lotion.
Normal saline.
Dextrose 5%.
Local anaesthetic injection.
Injection of heparin.
Contrast in bowls containing hot water.
Extra drugs for sedation.
Injection of anti-histamine and hydrocortisone.
A selection of indwelling intravenous needles.
A selection of intramuscular needles.
A selection of syringes.
Sterile gauze squares (swabs 4″ × 4″).
Sterile lifting (or Cheatle) forceps.
Ampoule files.
3″ Elastoplast.
Small water-proof dressings.
Writing pad and pencil (to record drugs and contrast given).
Masks and theatre hats.

Also needed

A good spot light.
A lined container for refuse.
Fenwall pressure bag (Baxter laboratories) and drip stand
(It is assumed that equipment and drugs for resuscitation will be readily available.)

Simple methods in vascular radiography

1. **'Single-Shot' films.**—A. The shoulder region (on a 12″ × 10″ cassette without a grid) or the arm (diagonally across a 17″ × 14″ cassette with the tube raised—no grid) can each be shown completely during axillary, or the less commonly used subclavian arteriography. 15 ml. of contrast are used for the shoulder, exposing just before the end of injection. To show the whole arm 25–30 ml. are injected, the exposure being made just after injection ends.

B. Aortography similarly can be carried out on a Bucky couch. The first film is exposed towards the end of the injection and is changed as quickly as possible and a second exposure made. When the aortogram is for occlusive vascular disease one can show the leg vessels in the following way[1]:—

The first film is exposed over the aortic bifurcation and iliac vessels; the tube and linked Bucky tray are then moved quickly to the knee region, changing the film *en route*. The second film is exposed as soon as possible. To show the intervening thigh and the calf a second

[1] Acknowledgements are made to Dr. F. Starer who described this method to us.

injection is given making the first exposure over the thigh soon after injection has ended and the next exposure over the calf. This method may also be used for femoral arteriography.

C. Alternatively there are the single exposure/single injection techniques. These use either a very long cassette or overlapping cassettes—one in the Bucky tray and one with a grid on the table top. The X-ray tube is raised to its full height to cover the whole leg. 35–40 ml. of contrast are used to fill the whole arterial system of the leg before exposing. Films or screens of differing speed or a suitable filter under the tube are needed to produce comparable densities in calf and thigh regions. We have no personal experience of this method.

D. Yet another possible method of obtaining films without special equipment is to employ the long exposure technique. This involves making a three-second exposure, beginning during the injection of the contrast. A second, similar exposure is made during the nephrographic phase as soon as the cassette has been changed. The preliminary film is also made with a long exposure. The resulting films can be of excellent quality but the technique is not suitable for use on a patient who cannot remain absolutely still and apnoeic. For practical purposes, therefore, general anaesthesia is needed.

2. **Multiple double-wrapped films.**—For the hand it is possible and most desirable to use non-screen films instead of cassettes. The exposure time should be kept relatively short (·05 second or less) by using a high mA. A simple cardboard, plastic or wooden spacer holds the hand just off the table and provides a space into which the films can be slid. The exchange is then made by pulling one film out and sliding in the next. A simple hand changer is easily constructed for departments which perform frequent arteriographic examination of the hands.

3. **The Tunnel.**—This consists in essence of a pair of boards with a gap between them through which the cassettes can be slid. The gap is maintained by wooden spacers. Grids are fastened to the upper board with adhesive tape and the patient lies upon it with the area to be examined above the gap. Two assistants are needed, one to push in cassettes, the other to catch the displaced cassettes as they emerge on the other side.[1] The levels at which successive cassettes are to be pushed in are marked in chalk on the side of the tunnel beforehand. This technique requires considerable practice and robust cassettes, but if the same assistants are regularly available it can be almost as satisfactory as a proper changer. With unpractised assistants the possible mishaps are numerous. The cassettes awaiting insertion and those which have been exposed are kept as far from the field of irradiation as possible. With this method as in other vascular examinations 'matching' cassettes should be used.

[1] Alternatively a padded lead-lined box may be attached so as to catch the emerging cassettes.

CHAPTER 26

ARTERIAL AND VENOUS PUNCTURE
AND CATHETERISATION

IN our first edition a distinction was made between peripheral arterio-graphy using needle puncture and Seldinger catheterisation which was usually employed to demonstrate the aorta and its main branches. In current practice this distinction is less firmly drawn. Catheterisation is not uncommonly employed for arteriography of peripheral vessels and when direct puncture of an artery is the means for injecting contrast it should, we believe, be carried out with an arterial cannula, so providing a stable puncture. As the same cannula is employed for introduction of the Seldinger guide wire the techniques of puncture for cannulation and catheterisation will be described first, together, with points in technique common to both procedures. In later chapters the use of these methods will be discussed in more detail.

Equipment used:—

1. **The arterial cannula.**—The recommended cannula is a modified Sheldon arterial cannula (MD). This is a thin-walled 18 gauge cannula with a needle-shaped stilet. The thin wall allows a lumen large enough to accept a PE 160 (O.D. 0·035″) guide-wire and so it can be used for puncture before Seldinger catheterisation or for direct contrast injection. The 17 gauge Seldinger needle-cannula is similar in principle but being larger, is more traumatic. Simple short bevel needles can, of course, be employed but are less satisfactory and are not recommended. The Longdwel Teflon needle-catheters are an alternative with advantages similar to those of the arterial cannula but we have no personal experience of their use for arterial work.

2. **Polyvinyl connections.**—These have three functions in arteriography:

(*a*) To prevent movement of the syringe from being transmitted to the cannula.

(*b*) To keep the operator's hands away from the field of irradiation.

(*c*) To allow the flow-back of blood from the cannula to be easily observed.

They consist of a male and a female adaptor linked by a length of polyvinyl tubing (translucent vinyl no. 4 (Shore 80)), (Fig. 29). The adaptors may be Luer-Lok or plain Luer as required.

The common sites for peripheral arterial puncture are the femoral, axillary and brachial arteries. There are many aspects of arteriographic technique which are similar in all these regions as well as in arterial punc-

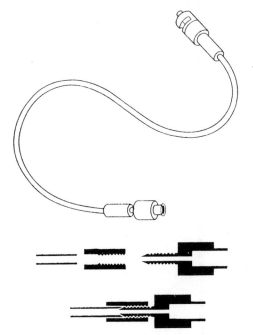

FIG. 29.—Polyvinyl connections for use in arteriography. The
assembled connection is shown above; the component parts
are shown below.

ture for Seldinger catheterisation. The points which these procedures have
in common will therefore be discussed first and details of examination of
particular areas dealt with subsequently.

THE TECHNIQUE OF ARTERIAL CANNULATION AND CATHETERISATION

1. **Preliminary assessment of the distal vessels.**—Before puncturing a
peripheral artery such as the femoral or axillary it is advisable to feel the
distal pulses in the limb. Their distribution and force are noted and
subsequently, before a catheter is withdrawn, the continued presence of
the pulses can be confirmed, showing that no damage has resulted from
the catheterisation. The measures recommended with impaired or absent
pulses after catheterisation are discussed later.

2. **Palpating the Artery.**—It is usually easy to assess the exact position
of the more superficial arteries. The deeper arteries, particularly the
femoral, may also be easy to feel but their pulsation often diffuses over a
wider area than that of the vessel itself. The fingers should be eased to and
fro across the line of the artery until its precise position is certain.

When it is difficult to feel a deep artery the palpating fingers (chiefly
the left index and middle) may become tired and their sensation blunted.
At such times one should palpate intermittently and use bimanual palpa-

tion. It should be noted that effective premedication or general anaesthesia, by lowering the blood pressure, may increase the difficulty of palpation.

3. **Local Anaesthetic.**—(10 ml. syringe with 10 ml. of 1% lignocaine.) A small skin wheal is raised at an appropriate point in the line of the artery, usually distal to the site of intended arterial puncture. An intra-muscular

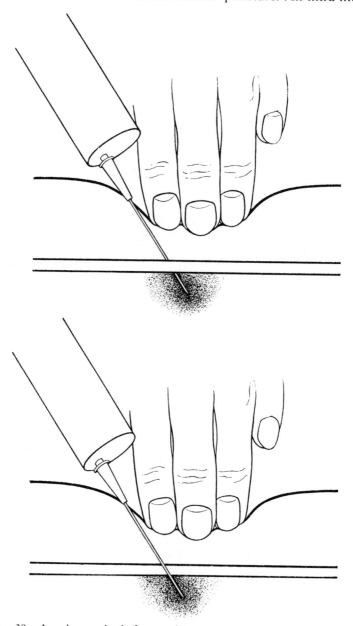

Fig. 30.—Local anaesthesia for arteriography. After injection of a small quantity of local anaesthetic into the skin, the main injection is made *deep* to the artery, *a.* first on one side, *b.* then on the other.

needle (hypodermic No. 0 or 1) is then attached to the syringe and directed obliquely down towards the artery. A small quantity of local anaesthetic— not more than 1 ml.—is injected superficial to the artery; the needle is then thrust in just to one side of the vessel, and about 4 ml. is injected *deep* to the artery. A further 4 ml. is injected similarly on the other side (Fig. 30). This lifts the artery towards the surface and avoids making it more difficult to feel. It also anaesthetises the sensitive underlying perio- steum when this is near the artery. Before each injection the plunger is drawn back since either the artery or its neighbouring vein may be punctured accidentally.

4. **Fixing the Artery.**—If an artery is mobile, as with the carotid or the brachial in mid-arm, it is very liable to slip from under one's fingers and away from a needle. To make puncture possible the artery must first be fixed by displacing it to one side and then pressing it down onto the bone behind. Once this technique has been mastered puncture is much easier.

5. **Puncturing the Artery.**—The skin at the site of the local anaesthetic wheal is nicked with a small sharp scalpel to facilitate entry of the arterial cannula. If Seldinger catheterisation is to be undertaken the guide-wire, catheter and perfusing fluid are made ready. Before simple arteriography an assembly consisting of a syringe of saline,[1] a two-way tap and an arteriographic connection is prepared and the saline is brought to the tip of the connection, displacing all bubbles. A short (15 cm.) guide-wire should also be available with an attached needle-stop. The arterial cannula is taken in the right hand and advanced into the subcutaneous tissues at an angle of about 45°; too high an angle makes for an unstable puncture, too low an angle makes puncture more difficult. The exact line of the artery is felt and, maintaining this palpation and, if necessary fixing the artery, the assembled arterial cannula is thrust firmly down through it. The central stilet is removed leaving the cannula in the tissues. The cannula is then slowly and steadily withdrawn. If it has correctly trans- fixed[2] the artery there will be a slight quiver and blood will spurt out with a characteristic arterial pulse. If the arterial cannula is not in the artery[3] the position of the vessel is carefully re-assessed and the puncture is again attempted. It is important to recognise the vigorous spurt from a cannula properly positioned in the lumen. A subdued or 'damped' flow may result

---

[1] For convenience the term 'saline' is used in many places during these accounts though, in fact, as explained later in this chapter, 5% dextrose solution is often a more suitable fluid.

[2] With experience it may become possible to enter the lumen directly but this is not always practicable. The transfixion method is easier and in our hands has proved satisfactory and safe.

[3] If a simple needle is used instead of the arterial cannula or catheter recommended, it may become dislodged from the artery with weakening or cessation of the flow and with or without haematoma formation. Should this happen the following manoeuvre may re-establish it in the lumen:—

The needle is thrust deeply through the artery. The tubing is disconnected and the stilet is inserted fully into the needle twice, wiping it between. The tubing is then *firmly* re-connected and 1 ml. of saline briskly injected. Then the syringe is taken off and the needle steadily drawn out until the flow starts again, as it usually does. This is particularly useful when a haematoma has formed.

Should the manoeuvre fail and re-puncture be necessary through a haematoma it is best to press firmly on the puncture site for 5–10 minutes; this stops further bleeding and helps to make the artery easier to feel.

from a needle or cannula whose tip is only partly in the lumen. This requires further withdrawal until an adequate pulsating flow is obtained.

When satisfactory puncture has been achieved the technique varies according to whether cannulation or catheterisation are to be undertaken:

Arterial cannulation.—The butt of the cannula is lowered to the skin

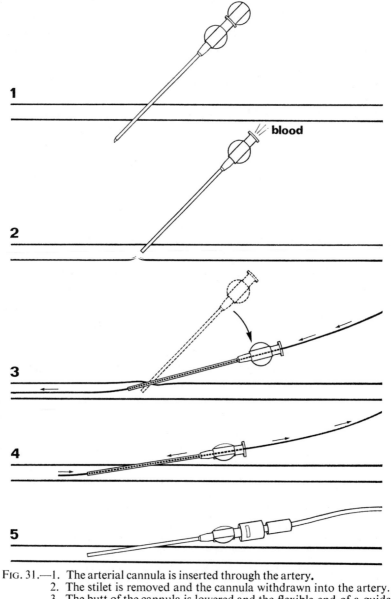

Fig. 31.—1. The arterial cannula is inserted through the artery.
2. The stilet is removed and the cannula withdrawn into the artery.
3. The butt of the cannula is lowered and the flexible end of a guide introduced into the arterial lumen.
4. The cannula is advanced into the artery and the guide is withdrawn.
5. An arterial connection is attached and flushing commenced.

so as to direct it along the line of the artery and a short guide-wire (with an attached needle-stop) is advanced gently up the artery *flexible end first*. The cannula is then advanced over the guide another 2–3 cm. so as to secure its position in the lumen. The guide-wire is then withdrawn. Once this has been done the chance of accidental dislodgement of the cannula is very considerably reduced. (Fig. 31).

Seldinger catheterisation.—The butt of the cannula is lowered and the flexible end of a guide-wire advanced 15–20 cm. into the artery the flexible end leading. The cannula is then withdrawn, and after wiping the guide-wire clean of blood the catheter is slid over the guide and into the artery. The procedure is described in more detail in Chapter 29.

6. **Maintaining patency of the 'needle'.**—Whether needle, cannula or catheter are used the lumen must be kept patent by instilling fluid, usually either saline or dextrose. Perfusion can be either by manual injection, with a sterile tap to prevent flow-back when the syringe is changed, or by a pressure drip. The latter is preferable (1) because the operator is freed from the need to instil fluid continuously, (2) because when an open jug of fluid is used as a source for injection cotton dust, room dust and talcum powder inevitably contaminate the fluid and may be injected.

The Steriflex system (A.H.) allows pressure infusion without risk of air-embolism when the Fenwall pressure bag (Baxter) is used in conjunction with it. Alternatively a simple drip bottle may be employed. When used with air-pressure injection the hazard of air-embolism cannot be avoided; an alternative method which overcomes this difficulty is to elevate the drip bottle to the highest possible point so that gravity is then available to overcome the arterial pressure.

Normal saline is commonly used for infusion but in patients with hypertension, renal disease, etc., an excess of sodium is undesirable. There is, therefore, some advantage in using dextrose 5% as a routine *except in diabetic subjects*; (see also page 220 concerning the use of dextrose solutions in aortography). Heparinisation of the fluid by addition of 500 i.u. (5 mg.) per 500 ml. of infusing fluid is sometimes recommended. It is uncertain whether this has any significant effect. The dose employed does not produce a general effect on blood clotting and it seems probable that the thrombi encountered in relation to catheters are platelet thrombi; their formation is unaffected by heparinisation.

7. **Test injection.**—When employing a needle for arteriography a brisk injection of 10 ml. dextrose solution may be made before injection of contrast. A normal flow-back after this reassures one that the final injection will not dislodge the needle.

8. **Injection of contrast.**—Hand injection through glass syringes always involves the risk of their breaking and now that plastic disposable syringes are available their use is desirable for all types of arteriography when manual injection is made. Plastic syringes are extremely robust and the plunger rarely 'sticks' to the cylinder during injection. They are not, of course, suitable for pressure injection when mechanical methods are used.

Glass syringes should be used only in conjunction with metal guards and a 'mushroom' head (M.D.) on the plunger to facilitate injection. The contrast is injected hot, but not unbearably so. The tubing is preferably attached to the syringe by a Luer-Lok connection. As a rule the first film should include the puncture site so that one must always call for the first exposure before the end of the injection. The exception to this occurs when the area of interest lies away from the site of puncture and all the films are taken in one particular area.

9. **Calling for Exposures.**—Usually the operator calls 'take' or 'shoot' for each exposure. The radiographer who is making the exposure calls 'pull' immediately after she has exposed. This tells the assistants who may be moving the tube and changing the cassettes to act at once. All calls should be loud and clear. It is most important that the various manoeuvres including moving the tube, cassette changing and exposure alteration should be repeatedly practised until they have been perfected. This should be completed before the patient enters the room.

10. **Removal of the catheter or needle.**—Before a catheter is removed from the artery it is advisable to withdraw it so that its tip lies close to the point of puncture but is still in the lumen. The distal pulse (usually the dorsalis pedis) is then felt to make sure that it is still present and of comparable strength to the pulse felt before the examination began. If the pulse is present the catheter may then be fully withdrawn and compression of the area of puncture is begun by pressing swabs firmly down onto the puncture site with the palm of the hand or with the fingers. The suggested routine if the pulses are diminished or absent is outlined in the following section. Bimanual pressure of the puncture site is less tiring.[1] The subclavian and axillary arteries are often difficult to compress effectively but compression must be attempted. Puncture sites in the carotid, axillary or brachial arteries are compressed for not less than 5 minutes and in the

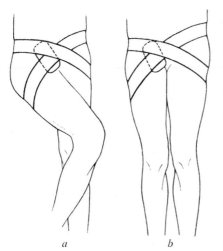

FIG. 32.—Application of strapping after femoral artery puncture. *a.* A reasonable length of Elastoplast is applied over a thick pad of gauze swabs, with the hip flexed and adducted. When the leg is straightened (*b*), firm pressure is produced over the puncture site.

Alternatively the Elastoplast may be applied with *firm* pressure on a thick pad.

[1] An alternative method suggested by Dr. Roland Terry is as follows:—
A used Steriflex infusion bag (see page 199) is filled with about 500 ml. of water and a knot is tied in the tubing by which it fills and empties. This provides a firm but flexible pressure bag which is placed over the puncture site and forced down onto it by means of a Bucky band. This technique is not tiring and the control of pressure is very precise, but it is not possible if the arteriographic table does not have suitable attachments for a Bucky band.

femoral artery for at least 10 minutes and longer in hypertensives or after prolonged manipulation of catheters or use of wide-bore catheters. All sites must be carefully watched for a minute after release of pressure and gently palpated to make sure no haematoma is forming. Injection of 1000 units of hyalase in 10 ml. water for injection helps to disperse any significant haematoma but is only used when all bleeding has stopped. After puncture of smaller arteries such as the subclavian, axillary or brachial a light dressing is usually sufficient provided no haematoma formation is observed. This may also apply to the larger arteries but when a haematoma forms at any site or after repeated catheter manipulation a pressure dressing is useful. For the brachial, axillary or femoral arteries a large pad of gauze swabs is firmly strapped over the puncture site. After femoral artery puncture the hip is semi-flexed and slightly adducted and Elastoplast strapped tightly across as in Fig. 32. This produces firm pressure when the hip is straightened. Even with this firm strapping the puncture site should be seen again *by the operator* when the patient returns to the ward, especially if hypertension is present, since rapid haematoma formation may result from the movements on and off the trolley on the journey back. The disadvantage of this type of strapping is that it can produce considerable trauma to the skin.

After femoral arteriograms, as after puncture for catheterisation, the patient is best restricted to bed for the following 24 hours and the staff should be instructed to watch the groin for haematoma formation and to check the foot pulse on the appropriate side. All patients are admitted to the ward for the night following arteriography. The foot pulse (or radial pulse after subclavian, axillary or brachial puncture) is recorded quarter-hourly or half-hourly and the blood pressure half-hourly for 4 hours. Thereafter four-hourly recordings of blood pressure and pulse are made.

11. Anticoagulants.—Before performing an arteriogram one should check whether the patient is being treated with anticoagulants. It is preferable not to perform arteriography during such therapy but there are risks in sudden cessation of treatment and in practice there is seldom any difficulty with abnormal bleeding after arterial puncture or catheterisation provided the prothrombin time is not more than twice normal (J. Dow, personal communication). This is presumably because bleeding stops initially not by clotting but by occlusion of the puncture by platelet thrombi. Before direct puncture lumbar aortography where pressure on the bleeding point is not possible the clinical indications for the use of anticoagulants should be reviewed; even here, in a case of urgency, it is possible to undertake puncture of the aorta without necessarily incurring severe haematoma formation if the prothombin time is not unduly prolonged as mentioned above.

12. Arterio-venous fistula.—When injecting contrast to demonstrate an arterio-venous fistula larger volumes and more concentrated contrast than are usually employed will be needed. Injection must be made as rapidly as possible using a catheter and a pressure injector.

13. Disappearance of the distal pulse.—If the distal pulse should disappear during the course of an examination—and as indicated earlier the pulse should always be checked before withdrawal of a needle, cannula or catheter—the following routine is suggested:

1. With the catheter tip or needle situated just above the point of puncture a slow injection of papaverine, 30–100 mg., well diluted, is made. The papaverine reduces arterial spasm and is frequently effective in restoring circulation. It is given slowly over about 5 minutes so as to allow its effect to extend distally if spasm is relieved. The other reason for slow injection is that ventricular tachycardia and other cardiac irregularities may ensue and the heart rate is carefully checked during its administration.
2. 5000–8000 i.u. of heparin diluted in saline are injected through the cannula or catheter. This is intended to prevent propagation of any thrombus by clot formation.
3. 500–1000 ml. of low molecular weight dextran (Rheomacrodex-Pharmacia) is injected.
4. In patients who have been examined under general anaesthesia it may be worthwhile attempting infiltration around the artery with local anaesthetic in the hope of relieving any possible spasm.
5. A suitable amount of contrast is injected and either watched on the intensifier screen or a film taken to give an indication of the state of the occluded vessel. If the artery remains occluded after these measures surgical intervention will be necessary. The needle or catheter is removed and the puncture site gently compressed while awaiting the surgeon. Even at this stage it is worthwhile continuing to check the distal pulse since spasm may be extremely slow to resolve.

14. The use of gloves in arterial work.—It is not generally realised that the use of sterile rubber gloves during arterial work does not significantly impair the ability to palpate arteries. Each radiologist has to decide the level of asepsis which he will try to maintain in such procedures but gloves should not be dismissed from consideration because of their supposed effect on sensation. Before undertaking a procedure the talc should be rinsed from them.

CHAPTER 27

PERIPHERAL ARTERIOGRAPHY

FEMORAL ARTERIOGRAM

Ward Preparation

1. Groin and pubis are shaved.
2. Nothing to eat or drink for the previous 6 hours.
3. An enema saponis is given 2 hours before (this leaves the abdomen clear if an aortogram should prove necessary).
4. The bladder is emptied just before the patient comes to the department.
5. Premedication is given; this is usually papaveretum 20 mg. and scopolamine 0·4 mg. given intramuscularly $\frac{3}{4}$ hour before the examination.

X-Ray Department Preliminaries

1. The patient is given a brief explanation of the procedure. Stress is laid on the need to keep still during the injection of contrast in spite of any discomfort or noise.
2. The pulses are felt and the lower abdomen auscultated in an attempt to localise any blocks or stenoses. This helps in timing the exposure if the test injection cannot be watched on a television monitor. If the femoral artery pulsation is weaker than the pulse in the other groin, a catheter should be passed up the femoral artery on the other side (i.e. the artery with the stronger pulsation). If the pulsation is definitely impaired on both sides one should consider whether direct puncture aortography may not be preferable. Whatever the mode of examination it should be remembered that the femoral artery is not an entity in itself and the aortic bifurcation should always be demonstrated in every femoral arteriogram: for this reason it is advisable to pass a catheter on the side under examination, taking, if necessary, a separate film of the iliac artery with the catheter tip at the bifurcation before undertaking the full 'run' down the leg. When the examination is made by injection at the aortic bifurcation via a catheter placed in the opposite artery and when it is known that there is no clinical interest in this artery the cuff inflated round the appropriate thigh may help to divert the contrast into the 'bad' side.[1]

[1] If a satisfactory puncture has been achieved on the symptomatic side prior to catheterisation and the catheter will only enter for a few centimetres a forceful manual injection may be given provided there is a free flow of fluid from the catheter and provided screening on the television monitor shows no evidence of extravasation or subintimal injection. On occasions, particularly if the Valsalva manoeuvre is used, the 'head' of contrast may reach the aortic bifurcation; care must be taken to ensure that the catheter tip is not caught under an intimal plaque or lodged in a minor branch artery.

3. The patient is given further intravenous sedation as required.

4. **Preliminary films.**—A series of overlapping exposures will be needed when examining the whole leg and these will extend from the aortic bifurcation to the mid calf. At least one preliminary film in the pelvis or thigh region will be required. Less experienced radiographers may prefer to have preliminary films of the whole series. The exposure is usually altered by varying the kV. A grid is needed for the pelvis and the thigh regions and in most devices it is an intrinsic part of the changer.

Other positions of the limb may be adopted when examining particular areas. For example the lateral leg position is useful for such lesions as popliteal aneurysms or tumours. Such positions cannot be adopted with safety if a needle alone is used and this is a further argument in favour of the use of a catheter for all femoral arteriograms where it is possible. Depending upon the problem under investigation the catheter is passed a variable distance into the femoral or iliac artery, remembering that a length of about 22 cm. is necessary to bring the tip just past the aortic bifurcation; it is then *strapped* to the skin with sterile adhesive strapping[1] (see Fig. 39). The lateral and A.P. foot positions are useful in studying circulation around the ankle joint and in the foot.

When a 'moving' series is to be taken and automatic movement is not available, the floor by the foot of the tube column is marked for the various tube positions. When a 'tunnel' is used the corresponding film positions are marked on its side. After the preliminary films have been viewed any necessary alterations in position or exposure are made. Other points to note are as follows:—

(a) The patient must lie on one side of the couch so that the leg lies over the centre of the film.

(b) The leg is held in full internal rotation using a crepe bandage with a clove hitch round the foot or a sandbag against it.
This opens out the gap between tibia and fibula, showing the anterior tibial and peroneal arteries more clearly. A pad under the knee helps make this position more comfortable. It is relaxed for the puncture and between exposures.

(c) It is again emphasised that moving the tube, film changing and exposure changing should all be repeatedly practised beforehand until they have been perfected.

Procedure

While the preliminary films are being processed the operator, wearing a lead apron and a mask, scrubs up and puts on a sterile gown, and also gloves when these are used. The nurse assisting, also scrubbed up, prepares the groin (see chapter 4) and drapes a waterproof paper sheet over the thighs.

---

[1] Small lengths of sterile autoclave tape may be prepared by rolling onto a glass cylinder such as the plunger of a syringe or a test tube. This tape retains its adhesiveness after autoclaving when prepared in this way.

The abdomen, opposite leg and couch are covered with sterile towels. The operator, if right handed, normally stands on the patient's right. The puncture should always be in a cranial direction[1] regardless of whether this is for catheterisation or direct needle puncture. The technique of puncture is described earlier (page 197). When puncturing the femoral artery it is advisable to aim for the superior pubic ramus. This allows the artery to be fixed against the underlying bone and prevents the puncture from being too deep. It also ensures that the puncture is within the thigh; any subsequent haematoma is then less likely to spread into the abdomen, scrotum or retroperitoneal tissues.

Before injecting contrast medium the patient must be warned 'there will be a feeling of heat in your leg; even if it's very uncomfortable you must not move. Keep quite still', etc., etc.[2] We usually inject 20–25 ml. of 65% Hypaque or Conray 280. When the artery is thought clinically to be normal, injection is made as rapidly as possible; with femoral artery blocks it should be rather slower.

Whole Limb Arteriography.—Here the first (pelvic) exposure is made when about 4–6 ml. are left in the syringe. Timing of later exposures varies with the state of the artery and should be governed by the earlier timing of the test injection of contrast or of radio-isotope. With a normal circulation successive exposures are made as soon as the tube can be moved or the patient moved over the changer, i.e. at about 2–3 second intervals. In the presence of a known femoral block one should inject rather more deliberately[3] and pause at the knee for a period governed by the earlier timing. The first upper calf exposure will then be made as soon as the tube reaches it and successive exposures after intervals of about 3–4 seconds. Should it be thought that the blocks are all in the calf vessels the knee exposure is made without a pause and calf exposures 4 and 9–10 seconds later. The lower margin of the calf film need not normally extend much below the mid-calf region since disease below this area is not accessible to operation and the main object of the film is to show the origin and filling of the three main calf vessels. It may be helpful to point out that gangrene of the toes usually means disease of the calf vessels as well as the femoral artery. In cases of arterial obstruction it is not always possible to gauge the rate of flow down the artery with complete success, and it may be necessary to repeat the injection and take more films of a particular area with different timing of the exposures. Sometimes without any apparent block the contrast in the calf vessels seems to 'fade away' and never reach the

[1] The advantages of retrograde puncture for femoral arteriography are: i. that it is easier; ii. that it is unimportant if the profunda femoris is punctured in error; iii. that by forced injection with an occluding cuff on the thigh it is often possible to show the iliac vessels—should this be desired. The main disadvantage of retrograde puncture is that the cannula and connection often overlie the femoral artery; a lesser drawback is that it is slightly more difficult to 'time' the passage of contrast down the leg. The former disadvantage can be overcome by using a catheter—yet another argument in favour of this practice.

[2] It should be noted that however attractive in theory, the prior injection of local anaesthetic into the lumen of the artery does not reduce pain.

[3] The presence of a femoral block, particularly a high block, usually adds considerably to the pain of the injection.

ankle. This is usually due to spasm and the injection should be repeated after intra-arterial injection of tolazoline (Priscol-Ciba) 25–50 mg. in 20 ml. of saline or after occluding the circulation by a thigh cuff for 5 minutes. This appearance may also be due to hypotension and this possibility should be checked before injecting tolazoline.

Local Arteriography.—The problems in timing are fewer when the whole film series are to be taken centring on one particular area. The first three films are taken in quick succession and the last about four seconds later. The timing of the initial exposure varies according to the region. For the lower femur it is made as the injection ends, for the upper tibia about a second later and for the ankle about four seconds after the injection. is completed.

AXILLARY ARTERIOGRAM

We carry out this examination under general anaesthesia because axillary puncture is sometimes difficult and because of the proximity of the brachial plexus. There is however no absolute contra-indication to local anaesthesia. Puncture of the axillary artery may be indicated for several reasons. Excluding cerebral angiography which will not be discussed, the chief indication for axillary arteriography is for the examination of the shoulder, arm and hand where, for some reason it is not possible to approach the upper limbs via the subclavian arteries from below (via the femorals.). The authors prefer the femoral approach since it is easier and safer carrying fewer complications and can be performed under local anaesthesia. The other indication for axillary puncture and catheterisation is for thoracic or lumbar aortography in those cases where an approach from below is precluded.

The film changer employed varies with the area under examination. For the shoulder and arm we have kept to the usual cassette changer or used single cassettes—see p. 192. For the hand and wrist we strongly recommend the use of non-screen films since they give better detail than cassettes. Being light the films are quite easily exchanged without special apparatus: if the patient's hand is placed on a perspex plate held 5–10 mm. off the table, the films can be slid in and out of the space below the plate, removing one and inserting the next. When frequent axillary or subclavian arteriograms are performed it is worth constructing a simple changer.

To give an elegant demonstration of the fingers they should be evenly separated and strapped to the surface of the changer. Sellotape is the best form of strapping as it does not show on the films.

Ward Preparation and Premedication.—These are as for any general anaesthetic. The axilla must be shaved. An anaesthetist is 'booked' for this examination.

Department Preliminaries

1. The patient is given a brief explanation of the procedure and warned that he may have a bruised armpit when he comes round.

2. He is moved onto the couch and preliminary films are taken before or after induction of the anaesthetic, as convenient. Care is taken to pad under the arm when it lies across the edge of the couch.

3. The arm is now raised to just above a right angle, the elbow is flexed and the hand rotated outward. The head is turned to face the opposite side. The hand is then tied or held by an assistant to keep the arm in position.

4. The axillary region is cleaned and draped so as to cover both the chest and the arm.

Procedure.—In the position adopted the artery will usually have been pushed forward by the head of the humerus but it may be necessary to alter the degree of rotation of the arm or its abduction or to insert a pad under the head of the humerus to make it more easily palpable. The aim is to bring the artery superficially at the point of greatest pulsation and to provide a rigid background. A skin nick is made after the vessel has been located and it is fixed against the head of the humerus. A needle, either a shortened 18 gauge needle or the thin-wall 18 gauge arterial cannula mentioned earlier, is introduced, aiming upwards into the axilla along the line of the vessel with the needle shaft at an angle of about 35° to the skin. The flow from a 'clean' puncture should be pulsating and considerable but not equivalent to that seen in femoral artery puncture. A short flexible leader should now be fed in for a distance of only 4–5 cm. and the arm brought towards the side of the patient. After removal of the needle or cannula a tapered O.P.P. 160 polythene catheter is introduced only far enough to ensure a stable position in the artery. A test injection of 2–3 ml. of contrast should be made to check free position in the vessel since accidental entry into branch vessels or collaterals is particularly liable to occur in this segment of the artery. For the arteriogram itself a hand injection of 15 ml. 280 Conray or 45% Hypaque is made. Timing of exposures may be difficult to assess for the forearm and hand if vascular spasm at the puncture site or below the wrist occurs. If digital vaso-spasm is suspected an intra-arterial injection of tolazoline (Priscol) 25 mg. in 20 ml. saline may be made before repeating the injection of contrast and taking further films. Normally for the hand the first exposure should be made about 3 seconds after completion of the injection and the remainder at intervals of 2 seconds. A second injection is frequently required to obtain a 'complete' vascular picture in all phases of circulation. It must be emphasised that the axillary artery is more intolerant of trauma than the femoral and as has been indicated, will often go into spasm when traumatised (especially in children and young women). Every effort should therefore be made to obtain a clean puncture at the first attempt.

After the catheter has been withdrawn from the vessel compression of the axilla must be maintained for 4 or 5 minutes and finally a large pressure pad applied with strapping. It is particularly difficult to detect haematoma formation or bleeding in this region because of the laxity of the tissues and the distance the blood may travel around the lateral chest wall without becoming clinically apparent. Haemostasis is particularly important in this region since a haematoma can press on and damage the brachial plexus.

Subclavian Arteriogram.—This has now become an almost obsolete approach to the arteries of the upper limb since most problems are tackled from the femoral or axillary arteries. General anaesthesia is again preferred for the reasons given under Axillary Arteriography and the remarks concerning film changing made under that heading also apply to this method of arteriography. Before puncture the preliminaries are as for axillary arteriography except that the head, neck and shoulder region are cleaned and draped.

Procedure.—The subclavian artery crosses the first rib just behind the insertion of the scalenus anterior. It lies posterior to the mid point of the clavicle and palpation should be in this region. The ease of palpation varies considerably and one may need to rotate the head or raise the shoulder to relax fascia and to move the clavicle. If the vessel cannot be palpated the operator presses with one finger tip at points in the anatomical course of the artery behind the clavicle while an assistant feels the radial pulse: when the latter disappears the position of the finger is the point of arterial puncture. A skin nick is made and the needle (a shortened 18 gauge) is introduced, angling its point laterally. (This is not always possible in bull-necked patients but it is desirable because it means that in this position the contrast is injected in the right direction and the needle lies more along the line of the artery, making the puncture more stable, and the butt easier to support.) The needle is thrust in, feeling for the first rib with the needle point; the artery is pinioned against the rib and a successful puncture is almost invariable. By directing the needle in this way it is ensured that if it passes beyond the artery it will strike the first rib or go lateral to it. If the initial attempt is not successful it may be necessary to make repeated thrusts moving the needle point gradually backward and forward,[1] but as long as it is coming down onto or lateral to the first rib the danger of pneumothorax is slight. One should remove and clean the needle after every three or four thrusts.

Once satisfactory puncture[2] is achieved the needle butt is supported on swabs or a surgical pack soaked in saline and the connection anchored with a towel clip. We use 15 ml. of 45% Hypaque or 280 Conray injected fairly rapidly but by no means at full speed. Should the puncture be unstable or the butt of the needle difficult to steady it is better for the operator to hold the needle while the assistant injects (or vice versa). Timing of exposures is often difficult to assess in this peripheral examination if fluoroscopic control is not available. For the hand the first exposure should be about four seconds after completion of injection and the remainder at two second intervals. Because of this difficulty we usually proceed to an immediate second series with different timing provided the flow-back is still good.

BRACHIAL ARTERIOGRAM

Ward Preparation.—As in 2, 4, and 5 of Femoral Arteriography.

The special points about this arteriogram are:—

1. It is usually under local anaesthetic, injected as described in Chapter 26.

2. The needle point is normally directed upwards whether for simple arteriography or for brachial artery puncture by the Seldinger method.

3. The brachial artery is punctured in the antecubital fossa or in the middle third of the arm. We prefer the former site because the danger of spasm is thought to be less here. When the mid-arm is used the artery must be fixed before puncture and this may be difficult.

4. A scalp vein needle with its attached tubing is ideal for puncturing this sensitive vessel and contrast can be introduced with sufficient speed to outline the vessels of the hand and fingers provided that a 19 or 20 gauge thin wall needle is used. Alternatively the usual arterial cannula is satisfactory. There is rarely any indication for catheterising this artery since the axillary is usually easier to puncture and certainly far safer.

---

[1] A less traumatic way of finding a difficult artery is to use an intramuscular needle on a syringe of saline.

[2] Occasionally there is doubt as to whether a vein or an artery has been punctured. The syringe is disconnected from the tubing which is raised well above the sternal angle. Venous pressure will be insufficient to fill the tubing.

5. The brachial artery is extremely sensitive and is well-known for its liability to spasm. Whenever possible, puncture should be at the first attempt; the first 'shot' is by far the easiest. This tendency to spasm makes the brachial an artery which should be punctured only by those with a considerable experience of arteriography. When injection of contrast is finished, it is advisable to inject tolazoline (Priscol) 25 mg. or papaverine (50 mg.) in 20 ml. saline slowly into the artery before withdrawing the needle: during brachial catheterisation warmed heparinised saline is instilled through the catheter and tolazoline is injected through the catheter as it is pulled out of the artery, because of the danger of spasm and thrombosis.

APPENDIX

Femoral arteriogram[1]

Ward instructions

The examination will be at .on. . . . .
Please:—
1. Shave both groins.
2. Give nothing to eat or drink for 6 hours before.
3. Give Dulcolax, 2 tablets the night before and an enema saponis 2 hours before the examination.
4. Ask the patient to empty the bladder before coming to the Department.
5. Obtain written consent.
6. Give . . . . .as premedication at . . . .

After-care

1. Keep patient in bed for 24 hours.
2. Quarter-hourly pulse and B.P. for 4 hours; then 4 hourly pulse and B.P.
3. Half-hourly inspection around dressing to ensure no haematoma formation is occurring.
4. Remove plaster dressing at 24 hours.

Trolley setting

Basic trolley setting (p. 191) with 2×9 cm. (18 S.W.G.) arteriographic cannulae (M.D.) and 15 cm. guide-wire O.D. 0·035″ (Selflex: P.P. or M.D.) with adjustable needle stop (B.D. no. 4185).

Axillary Arteriogram

As for femoral arteriogram except the axilla must be shaved and the bowel preparation omitted.
The patient must be prepared for general anaesthesia.

Brachial and Subclavian arteriograms

Ward preparation
1. If under general anaesthesia routine preparation for anaesthesia including obtaining consent.
2. If under local anaesthesia:—

Ward instructions

The examination will be at .on. . . . .
Please:—
1. Give nothing to eat or drink for 6 hours before. Obtain written consent.
2. Give . . . . .as premedication at . . . .

[1] This setting is suitable for simple femoral puncture but when catheterization is intended the setting for Seldinger catheterisation (p. 235) is applicable.

After-care

1. Inspect around dressing half-hourly for 6 hours.
2. Quarter-hourly radial pulse on side(s) of puncture for 4 hours; then 4-hourly T.P.R.
3. Remove dressing at 24 hours.

Trolley settings

Basic trolley setting with the following differences:

1. The needles used for subclavian puncture are shortened (6 cm.) 18 gauge arteriogram needles (M.D.). For brachial arteriography the 19G scalp-vein set may be employed, or the arteriographic cannula used.
2. 2–3 surgical packs are included for use (wet or dry) in supporting the butt of the needle if this is necessary.

It is advisable to have a sucker available for this as for any examination under general anaesthesia.

CHAPTER 28

DIRECT PUNCTURE LUMBAR AORTOGRAM

THE dangers of this procedure are well known but as Sutton (1962) points out, most of them arise because of faulty technique. If conducted carefully the risks are relatively few. The examination is normally carried out under general anaesthesia as it can be extremely uncomfortable, but local anaesthesia may be employed if circumstances make it necessary or desirable. We do not employ this method on children since it is technically more difficult and Seldinger catheterisation is usually possible.

Contra-indications
1. Patient on anticoagulants—bleeding diseases, but see p. 201.
2. Severe hypertension which should first be controlled.
3. Raised blood urea. In the presence of renal failure it is particularly advisable to ensure that the patient is not dehydrated and that the blood pressure does not fall unduly. The use of dextrose in 'protecting' the kidneys is mentioned below.

Ward preparation
1. The patient undergoes routine preparation for general anaesthesia with premedication as prescribed by the anaesthetist.
2. The bowels are prepared by Dulcolax tabs. 2 the night before and a Clysodrast or trickle enema 1 hour before the examination.
3. Blood urea estimation and routine urine testing should be part of the normal preparation. Any abnormality which might contra-indicate the examination will be discovered in time. Furthermore any renal impairment demonstrated at this stage will not later be blamed upon the aortogram.

N.B.—Outpatients are admitted for the night following this procedure.

Department preliminaries
1. Examine the patient. This is with particular reference to his pulses, to assess the site of any blocks. Auscultation over the iliac or renal arteries may reveal a systolic bruit indicating a possible stenosis. Palpation of the abdominal aorta may be possible and will show whether it lies in the usual position. Inspection of the back may reveal a lordosis or scoliosis.
2. Explanation of the procedure which includes warning the patient that his back will feel stiff and bruised for some days.

3. *Position.*—The patient lies prone for the examination and, if conscious, must therefore be well-premedicated and comfortable. The head is turned to one side on a pillow and the arms may be brought up round the pillow or may rest at the patient's sides. The ankles are supported on pads *to keep the toes off the table.* When the knees are to be included in the examination they should be drawn together. Before puncture the patient must be asked whether he is comfortable since restlessness later on is inconvenient and possibly dangerous. Kyphotic patients may need extra support under the chest. It may also be necessary to 'pad up' one side of the chest because turning the head to one side makes some patients lift their whole trunk.

N.B.—The patient should lie with his left side away from the tube column —otherwise the operator may find the column in his way (note the value of ceiling suspension).

4. *Preliminary films.*—The most important is that which shows the abdomen. The quality of film required is similar to that for a plain film or urography. It is better to have a film very slightly too dark than too light. Until a radiographer is experienced she may prefer to take films of thigh and knee as well where these are to be examined. Before the plain film is exposed markers (e.g. ampoule files) are put on the back at about D12 and L.3–4 level and the skin at these levels is marked indelibly. The area covered by the plain films depends on whether the examination is for kidneys or for vascular disease—in the latter case centring is lower. The film should cover at least 4–5″ above the site of intended aortic puncture; this allows for variation in placing of the needle and ensures that the test injection is fully demonstrated—it is imperative that the needle point be included on every 'test injection' film.

Contrast.—The volumes injected vary with the purpose of the examination. When the patient is under general anaesthesia larger volumes of more concentrated contrast are advisable since they make timing easier and improve the opacification of the areas being examined. In the following table the figures in brackets are those suggested for local anaesthesia as opposed to general anaesthesia.

Test injection—6–8 ml. 60% Urografin, 45% Hypaque or Conray 280.
Injection solely for kidneys—30 ml. Conray 420 (30 ml. of 60% Urografin, 45% Hypaque or Conray 280.)
    ,,   for aortic aneurysm or for leg arteries—40–50 ml. of Conray 420 or Hypaque 85%. (30–35 ml.)
    ,,   in suspected aortic occlusion—35 ml. of 65% Hypaque or Conray 420, injected rather more slowly than usual after instillation of 500–1000 ml. of 5% dextrose.

These figures refer to the volume reaching the aorta; when a connecting system with a large capacity is employed, allowance must be made for this.

Equipment.—The usual arteriogram connections and syringe are employed. The needle is 16 S.W.G., the standard length is 17–20 cm., the long needle is 22·5 cm. (This latter may be 15 S.W.G.) We do not use a mechanical injector, believing that hand injection is safer.

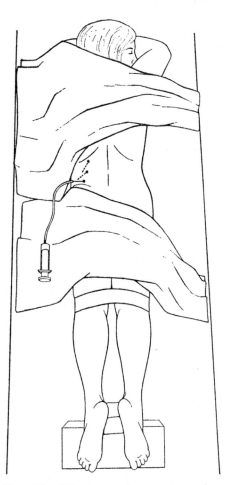

FIG. 33.—Direct puncture aortography. Showing the position of the patient and the dressing towels. The different puncture sites and the needle alignment for high, medium and low puncture are also shown.

Technique[1]

1. The usual aseptic precautions are observed. The patient's back is prepared in the usual way, taking care not to rub out the skin marks. The hips, thighs and shoulders are covered with sterile towels (Fig. 33). Sterile absorbent wadding is put on the couch to the left of the patient as blood or dextrose tend to be spilled here.

2. The plain film is inspected to determine the level of the skin markers in relation to the underlying vertebrae and kidneys. This enables one to select the point at which to aim the needle. *Either the plain film or earlier films of chest or abdomen may show the course of the lower thoracic aorta and so give an idea of its direction as it enters the abdomen.* Equally they may show aortic calcification or scoliosis, points which are also of value in assessing the likely position of the aorta and the possible presence of aneurysmal dilatation.

3. Medium-level puncture, i.e. for the renal arteries, is made about 9–10 cm. to the left of the spine and about 4 cm. above the iliac crest. After infiltrating the skin and deeper tissues with lignocaine 1%, the skin

---

[1] This description is based on the use of local anaesthesia. When a general anaesthetic is employed the needle is advanced with the stilet in situ.

H

is nicked with a small scalpel and the aortogram needle is introduced. It is directed about 45° from the vertical toward the spine and is aimed at the chosen level for puncture of the aorta, using the skin marks as a guide. The angulation is usually about 45° cranially but this varies, as discussed below. Further lignocaine is injected as the needle is advanced.

4. To begin with it may be helpful to direct the needle so that it strikes the edge of a lumbar vertebral body; it is then withdrawn 4–5 cm., directed a little more laterally and advanced. This should enable the needle to pass just lateral to and in front of the spine and thus into the aorta. With experience one acquires an 'eye' for the correct angulation and puncture can often be made directly, but even then, in any difficulty, the spine remains a most useful point of reference. Once the needle is aligned to clear the vertebral bodies another difficulty may result from striking a transverse process; the needle must then be withdrawn, angled more cranially or caudally, and advanced.

5. If no further obstruction is encountered the needle reaches the region of the aorta and it may be possible to feel the aorta pulsating at the needle tip. The remaining local anaesthetic is then injected, the syringe is detached and the stilet replaced in the needle. After this, short, 2–3 mm. 'stabbing' thrusts are made, removing the stilet after each advance to see whether the aorta has been entered. A gradual advance may slowly compress the aortic wall until the needle suddenly passes, not only through the nearer side but also through the opposite wall. This becomes apparent when the aortic lumen is entered by withdrawing the needle, so leaving a puncture in the opposite wall. It is in this situation that extravasation of contrast most commonly occurs. If the aorta can be entered directly, by means of the 'stabbing' technique described, extravasation of contrast becomes relatively uncommon.

When blood spurts back, indicating successful puncture, a Luer-Lok connection attached to a syringe of saline or dextrose is firmly fastened to the needle. The needle is then rotated through 360° in preparation for the test injection (see below). Before this is described further points relating to the technique of puncture will be discussed.

A. **The point of skin puncture and the angulation of the needle.**—The point of skin puncture is related to the intended level of aortic puncture. In high puncture the aorta is entered at about D.12 level, the needle entering the skin just below the 12th rib. For low puncture the needle is inserted about 3–4 cm. more laterally than for high puncture and just above the iliac crest. It is directed only slightly cranially, pointing almost transversely towards L.3. Medium-level puncture is usually aimed at L.1–2 level so as to inject contrast close to the renal arteries. In slender patients the needle is introduced more medially and in thick-set or lordotic patients more laterally than in average subjects.

Angulation ventrally may be increased a little in thick-set or lordotic patients because of the depth of the aorta in such subjects, (the longer needle is used in such cases). However, much of the necessary increase in depth of puncture is provided by puncturing more laterally and maintaining the usual angle of about 45°.

B. **The level of aortic puncture.**—The choice of the level of puncture depends in part on the purpose of the examination but the arguments for high and for low puncture should be weighed before choosing one or other as a routine.

High puncture.—This enters the aorta where it is widest and above the level of any major branches. The position of the aorta is more constant at this level and atheroma or aneurysm are much less common here. The drawback for demonstration of the leg vessels is that much contrast is lost into the visceral arteries and opacification of the leg vessels is therefore fainter. There is also a slight risk of damage to the spinal cord if a large dose of contrast is injected into the artery of Adamkiewicz.

Low puncture.—Enters the aorta where it is narrower and where it may be tortuous or the seat of atheroma or aneurysm. All the contrast injected, however, reaches the legs.

High puncture is mandatory in suspected aortic aneurysm or aortic thrombosis and is preferable when there is lower aortic calcification. It is also the site of choice for translumbar catheterisation (see below). Low puncture is preferable when the problem is one of vascular insufficiency of the legs and there is no reason to suspect abdominal aortic disease.

C. **Difficulty in puncture.**—Difficulties may arise for a number of reasons —unsuspected depth of the aorta, scoliosis, difficulty in passing transverse processes, tortuosity of the aorta or even aortic thrombosis. To some extent a blind procedure such as this is bound to be a matter of trial and error, guided by the basic principles of the technique.

Tortuous or thrombosed aorta.—These conditions may give rise to difficulty if low puncture is attempted so that when aortic tortuosity (particularly aortic aneurysm) or aortic thrombosis are suspected puncture should be higher. (Tortuosity of the aorta, like aortic aneurysm, usually begins below the renal arteries).

Transverse processes.—There may be surprising difficulty, real or imagined, in circumventing a transverse process. The reasons for this are:—

(i) When the needle is angled cranially the available gap between transverse processes is reduced.

(ii) Attempts at redirecting the needle are not successful; the needle is not withdrawn sufficiently and passes in along the same line as before.

(iii) What is assumed to be transverse process is in fact the vertebral body itself.

Bearing these points in mind it should be possible to guide the needle past, but if not it may be necessary to repuncture more laterally, or at a slightly different level.

Deep aorta.—The aortic depth is surprising in fat patients especially if they are lordotic; unless this is appreciated and a long needle used the aorta may never be reached.

Scoliosis.—A scoliosis convex to the left demands skin puncture further to the left; the tendency of the aorta is to move in front of the spine and the needle must be directed more transversely and deeper. Alternatively it may be necessary to puncture the aorta above the region of the scoliosis. A scoliosis convex to the right involves a puncture only slightly to the right of the usual site.

Low diaphragm in emphysema. This may mean that puncture at D.12 will sometimes involve traversing lung or pleura.

D. Other difficulties.—*Occlusion of the needle.*—When the needle has been pushed through the tissues a number of times it may become plugged with blood clot or muscle. Should this be suspected the stilet is passed down it once and $\frac{1}{2}$–1 ml. of saline injected further to clear it.

Pain as the needle is advanced.—When the examination is performed under local anaesthesia the needle seldom causes much distress when passing through muscle provided a smooth deliberate advance is maintained. Sharp pain usually results from striking sensitive periosteum or a nerve. The needle is withdrawn and the patient is asked to indicate the site of the pain without moving his body or hands. Redirection of the needle, with injection of local anaesthetic if needed, is usually adequate to stop pain. Occasionally patients will complain bitterly and move a great deal. It is commonly better to give further intravenous sedation since such patients may continue to move if sedation is inadequate, however much local anaesthetic is used.

E. Other ways of entering the aorta.—One does not always enter the aorta after feeling pulsation at the needle tip.

(i) It may happen that while still infiltrating down to the aorta one enters it unexpectedly. The syringe is disconnected and the connection and saline syringe are attached.

(ii) Sometimes a sharp advance of the needle after a slow advance may pass right through the aorta and this becomes evident on withdrawal. The tendency then is to lie too near the opposite wall and it is wise to withdraw a further 5 mm. before the test injection.

F. Entry into other vessels.—A flow-back of blood may be obtained when the needle is in the region of the aorta, but although arterial it may be very sluggish. This cannot be accepted; the flow from an aorta is sometimes only moderately forceful[1] but has a distinctly arterial character. A partial occlusion of the needle by clot may damp a normal flow but this will improve after instillation of saline; if not, a sharp injection of 3–4 ml. saline will normally clear it. Unless one is reasonably confident that the needle is in the aorta it is better to withdraw and try again.

It should not, however, be thought that branches of the aorta always give a poor flow. This is not so and part of the reason for the test injection is to ensure that the needle is not in a branch artery.

The test injection.—Once the needle is thought to be satisfactorily placed in the lumen of the aorta, it is slowly rotated through 360°, flushing continuously with saline or dextrose and observing the flowback into the connection. If the flow of blood from it stops at any stage the needle is not properly in the lumen and must be advanced or withdrawn (Stirling, 1957). This done, the manoeuvre is repeated. Once this preliminary has been successfully completed the test injection is performed. 6 ml. of 45% Hypaque are injected briskly, an exposure being made during injection. Alternatively the injection may be watched on the television monitor and if the needle is properly sited the contrast will be cleared at once after injection. If a small linear density remains visible on the screen this indicates a subintimal injection; the needle is then withdrawn 3–5 mm. and the test injection repeated carefully. Whichever method is employed to check on the appearances, *the test injection is the most important single factor in making direct puncture of the aorta a safe procedure*. Its purpose is:—

1. To show the level of puncture.
2. To show the position of the needle within the aorta, i.e. whether too much to one side or not.
3. To demonstrate possible subintimal or periarterial injection and hence to guard against a really large injection being made into these sites.
4. To show abnormalities of filling, e.g. all the contrast passing into the superior mesenteric artery.

The small volume makes timing of an exposure more difficult. The contrast is injected rapidly from a 10 ml. syringe and it is best to call 'Take' after 2–3 ml. have gone in. Pain from the test dose is unusual and if it occurs a second film is taken as soon as possible to show any evidence of periarterial contrast (subintimal injection of a test dose seldom causes pain). In the early days of one's experience this second film should be taken as a routine since it is not always easy to recognise a subintimal injection until some experience has been acquired. If all the contrast passes subintimally it is visible as a dense area spreading from the needle

---

[1] The flow-back from the aorta of a patient under general anaesthesia is usually much less forceful.

tip and following the wall of the aorta. The outline is usually very clearly defined and *the contrast persists in a second film*. More often only part of the contrast is injected subintimally, the remainder outlining the aorta in the normal fashion. The subintimal part may then be partly masked by the intra-aortic contrast but will show the same features of greater density (particularly if seen 'edge-on') and sharp outline (Fig. 34). It is important to look for this so that the needle position may be adjusted. Periarterial injection is usually easier to recognise (Fig. 35) since the contrast spreads out into the peri-aortic tissues and the patient feels pain. The density of the contrast is seldom as great as with subintimal injection and the flowback is more often impaired.

FIG. 34   FIG. 35   FIG. 36

FIG. 34.—Direct puncture aortography—subintimal injection. The majority of the contrast is localised around the tip of the needle giving an area of greater density, particularly where seen 'edge-on'. A little contrast outlines the aorta.

FIG. 35.—Direct puncture aortography—periarterial injection. The contrast extends into the para-aortic tissues. The patient will usually complain of pain.

Fig. 36.—Direct puncture aortography—satisfactory test injection. The 'jet' of contrast from the needle tip is helpful evidence that the needle lies away from the walls of the aorta.

The infusion of saline or dextrose is instituted from a drip-bottle or Steriflex container while the test film is developed. The appearances of a satisfactory test film are shown in Fig. 36. Note how the needle point lies towards the middle of the aortic lumen (in this projection) and note the 'jet' of contrast which is evidence that it is not too near the opposite wall.

On this basis any necessary advances or withdrawals of the needle are made. It may of course be necessary to withdraw entirely for any of the following reasons:—

1. Wrong siting of the needle in relation to the area to be shown, e.g. below the renals.
2. All the injection passing into a branch of the aorta.
3. Subintimal or periarterial injection (Figs. 34 and 35).

Provided that fluoroscopy or the test film show satisfactory appearances, the full injection and 'run' may be undertaken after an attempt has been made to time the rate of flow down the legs. This is usually by timing the arrival of contrast at the knees on the intensifier screen or by injection of radio-isotope, using counters over the knee. After isotope injections the isotope-containing fluid in the arteriographic tubing is withdrawn into the syringe and this is then returned to the Isotope Department or Physics Department for disposal.

Injection of contrast is made by hand except after translumbar catheterisation. A guide to the amount to inject is given earlier. In suspected aortic thrombosis it is particularly important to protect kidneys from the effects of contrast by prior administration of dextrose 500–1,000 ml. so as to ensure a diuresis (see below).

Timing of exposures

Abdominal aortogram.—The normal practice is to take 3 films in quick succession, exposing the first towards the end of injection, with a fourth film after a pause of 2–4 seconds.

For arterial disease.—For iliac blocks and/or femoral artery blocks a series of overlapping films is taken to show successively[1]:—

   (i) Abdominal aorta and iliac arteries.
  (ii) Iliac and upper femoral arteries.
 (iii) Femoral and upper calf vessels.
 (iv) Calf vessels if desired—if not a second exposure at position (iii) is made. Each film is exposed as soon as the tube and changer have been moved to the new position. A second injection and 'run' is often needed to demonstrate particular areas especially when there is disparity in rate of flow in the two legs. When it proves impossible to show the leg arteries satisfactorily on one side (usually because multiple blocks make timing difficult) it may be better to turn the patient over and perform a femoral puncture. This is not always advisable and, if the patient's condition is poor, femoral arteriography should be carried out at a later date. Should it be decided that an immediate femoral arteriogram is justified, after a second aortic injection and run, the needle is pulled out and the patient prepared for femoral puncture. If this second series of films is unsatisfactory retrograde femoral arteriography is then undertaken.

For aortic thrombosis it is better to centre the first three films on the abdomen and pelvis since it is most important for the surgeon to know the condition of the iliac arteries. The fourth film may be centred on the femoral arteries. In suspected disease of the visceral arteries a 'shoot-through' lateral is of value in showing their origins.

Repeat injections.—Provided that a diuresis has been induced there is little risk to the kidneys in repeat injections up to a total of three. Otherwise, for injection above the kidneys, not more than two injections should

[1] The series employed varies with the site of puncture and the film size of the changer. See also footnote [1] on page 204.

be permitted. Below the renal arteries, up to four injections may be permitted if a diuresis has been induced.

Difficulties.—Most of the difficulties encountered have been discussed already. This is a procedure where experience rapidly improves one's facility but at all times careful adherence to the principles of finding the aorta will help towards achieving good results.

Complications

Some haemorrhage from the puncture in the aortic wall is inevitable and occasionally this is severe enough to demand transfusion. Apart from this the complications of the procedure arise chiefly from:—

(i) Toxic effects of contrast media especially if a large dose is injected into a branch vessel.
(ii) Dissection of the aortic wall by contrast.

These complications are discussed fully by Sutton (1962) and space does not allow us to deal with them here. One point in particular may be made. It is known (Owen *et al.*, 1953) that any insult offered to a kidney is better tolerated when it is in a diuretic phase than during oliguria. Furthermore it has been shown in animal experiments (Morris *et al.*, 1956; Lance and Killen, 1960) that damage to the kidneys from intra-aortic injection of contrast can be prevented or considerably reduced by hyperhydration. It is therefore logical, when performing an aortogram, to reverse the state of oliguria produced by fluid deprivation and premedication. This is particularly desirable in cases where the danger of renal injury is increased, e.g. in aortic thrombosis, polycystic disease or hydronephrosis, but there is a strong case for making it a routine practice. 500 ml. of 5% dextrose[1] solution may be rapidly instilled while perfusing the needle after its introduction and before the full injection of contrast. Dextrose is instilled slowly during the remainder of the examination. A similar approach is used in catheter aortography.

After-care.—When the needle has been pulled out the puncture site is covered with a small dressing or with collodion. The patient is confined to bed for the 24 hours after the examination with $\frac{1}{4}$-hourly pulse and $\frac{1}{2}$-hourly blood pressure recordings for 4 hours followed by 4 hourly temperature, pulse and respiration recordings. A small number of patients complain of severe pleuritic pain after high puncture. This normally responds to reassurance and analgesics. A pleural effusion may be present for a few days.

Alternative technique—translumbar catheter aortography.—This method has been described by Amplatz (1963). The technique employed resembles the standard method in most respects. Instead of the usual needle, an

[1] Mannitol is the substance used for producing an osmotic diuresis in experimental work and it is undoubtedly more effective. It is, however, less readily available and simple dextrose solution is probably adequate. Miller *et al.* (1962) suggest 5% dextrose solution with 4% urea given by intravenous drip.

8 inch needle-catheter is employed for the puncture. This has a 16 gauge Teflon catheter drawn out over an 18 gauge needle and stilet. Puncture of the aorta is made at D.12 and the inner stilet is removed from the needle to confirm that successful puncture has been made. The needle is then withdrawn leaving the catheter in the aortic lumen. Provided a normal flowback is obtained a J-guide is passed up the catheter and about 10 cm. into the aorta. The catheter is then advanced over the leader and 5–10 cm. up from the site of puncture. The guide-wire is then removed and a flexible Teflon connection is used to link the catheter to a syringe of contrast (all bubbles being expelled from the connection). A test injection of contrast is watched on the screen to ensure that the catheter is satisfactorily placed. Sterile autoclave tape is then used to fasten it to the skin and dextrose is infused through the catheter while preparations are made for the series of films. The advantage of the method is that provided a flexible connection strong enough to withstand pressure injection is employed, a pressure injector can be used thus enabling the operator to stand clear of the field of irradiation and allowing a more rapid injection of contrast.

If difficulty is experienced when the guide reaches the tip of the catheter the latter should be withdrawn slightly because it is usually found to be too near the opposite wall of the aorta. A simple PE 160 guide-wire can be used instead of the J-guide but it is more likely to catch in any puncture in the opposite wall and the use of the J-guide is preferable.

References

AMPLATZ, K. (1963). *Radiology*, **81**, 927.
LANCE, E. M., and KILLEN, D. A. (1960). *Surg. Forum*, **11**, 139.
MILLER, H. C., WAX, S. H., and McDONALD, D. F. (1962). *J. Urol.*, **88**, 160.
MORRIS, G. C., CRAWFORD, E. S., BEALL, A. C. Jr., and MOYER, J. H. (1956). *Surg. Forum*, **7**, 319.
OWEN, K., DESAUTELS, R., and WALTER, C. W. (1953). *Surg. Forum*, **4**, 459.
STIRLING, W. B. (1957). *Br. med. J.*, i, 402.
SUTTON, D. (1962). *Arteriography*. (E. & S. Livingstone, Edinburgh).

APPENDIX

Ward instructions

The examination will be at .... on....................
Please:—
1. Shave back (if necessary) and pubis.[1]
2. Obtain blood group.
3. Give Dulcolax, 2 tablets the night before examination and administer a Clysodrast enema 1–1½ hours before examination.
4. Give nothing to eat or drink for 6 hours before examination.
5. Ask the patient to empty the bladder immediately before coming to the department.
6. Obtain written consent.
7. Give .... as premedication at....................

After-care

1. Keep in bed for 24 hours.
2. Quarter-hourly pulse and half-hourly blood pressure for 4 hours, then 4-hourly T.P.R.

[1] A pubic shave is not necessary for direct puncture lumbar aortogram itself but is usually included in the preparation in case either catheter aortography or femoral arteriography is required.

Trolley setting

Basic vascular trolley setting but with the following additions:

2 long aortogram needles (16 S.W.G.) (M.D.).

2 standard aortogram needles (16 S.W.G.) (M.D.).

For translumbar catheterisation an 8″ Teflon catheter-needle 01·0052 16T (catalogue 6735) (Becton Dickinson—G.U.) and a J-guide are added to the setting. A 7 F.G. Teflon connection (G.U.) is needed to allow pressure injection. It is made up as follows:

30 cms. Teflon tubing 7 F.G. (Genito-Urinary Co.).

Both ends flanged; (place adaptor collars on tubing before flanging).

2 collars of Kifa No. 1 tubing adaptor.

2 taps of Kifa No. 3 adaptors.

Lower shelf

Include ampoules of 45 % Hypaque or 60 % Urografin for test doses.

CHAPTER 29

SELDINGER CATHETERISATION

Part I

THIS technique (Seldinger, 1953), once mastered, may be applied to the introduction of catheters of various types into both arteries and veins. The technique of local anaesthesia, arterial puncture and kindred topics have been discussed in Chapter 26 which should be read before studying the present chapter. Femoral artery catheterisation will be described in detail as it is the commonest application of the method but the fundamentals of the technique are the same whatever the vessel to be catheterised. The advantages of aortography by this route over direct puncture aortogram include the ease of placing the injection at the chosen level, the greater scope for moving and positioning the patient with safety, and the ability to compress the point of puncture.

Contra-indications

1. *Aortic thrombosis* precludes the use of this method from the femoral arteries. *Iliac thrombosis* prevents its use on the side of the obstruction.

2. *Coarctation*[1] only precludes its use if the pulses are entirely impalpable, but even if catheterisation is successful the stenosis may prevent the catheter from reaching the aortic arch.

3. *Iliac stenosis or kinking.* It is often possible to pass a catheter through a narrowed or tortuous area but this may be difficult and the likelihood of damage to the artery is increased. For this reason, when there is clinical evidence of iliac narrowing (impaired femoral pulses; bruit over the iliac arteries) an alternative route should be considered (e.g. direct puncture aortogram, axillary artery catheterisation. Age is not in itself a contra-indication.

4. *Bleeding diseases or patient on anticoagulants.* The importance of this as a contra-indication depends on the severity of the bleeding tendency. If it is mild this route is preferable to those involving puncture of a vessel which cannot be compressed, so that when the indication for aortography is strong these are not absolute contra-indications as they can be for direct puncture aortography.

Ward preparation.—1. This is similar to that for direct puncture aortography (p. 211). The colon and bladder must be empty. *Both* groins must be shaved. Blood urea estimation and routine urine testing should be

[1] The most valuable approach for demonstration of a coarctation is via the right axillary artery.

carried out. Premedication is required; the type used varies according to whether a general anaesthetic is to be given. The blood group should be obtained.

2. It is desirable to see the patient in the ward before the examination so that the presence of normal femoral pulses can be confirmed and the examination explained. It is worth while asking for a history of strokes and auscultating the carotid arteries; a fall in blood pressure during an aortogram may lead to a hemiplegia in patients with an impaired cerebral blood supply.

Department preparation.—The preparation of catheters is discussed in the Appendix. This can be carried out at the time of the aortogram but it is much more convenient to prepare catheters in batches well before the examination. Alternatively pre-formed sterilised catheters are now available (see Appendix to part 2 of this chapter).

Department preliminaries.—1. Palpation of femoral arteries, with auscultation over the iliac arteries if both femoral pulses are reduced. Any diminution of the pulse on one side is an indication to use the other side. When both sides are affected one should consider whether direct puncture may not be preferable; if not the side least affected is used. *The pulses of the feet should also be felt*, particularly on the side to be punctured. Any impairment which may later develop can then be correctly attributed to the puncture and appropriate therapeutic measures taken (p. 202).

2. The procedure is explained, warning the patient about the warmth of the injection and the need to keep still and hold the breath when asked. He is also warned that there will be a tender 'stiff' groin with an area of bruising for some days.

3. A preliminary film is taken. For the kidneys this normally includes both the kidneys and the pelvis but this will vary with the clinical problem and the site under examination. The centring point and edges of the area covered should be marked on the skin so that if the patient moves he can be repositioned without difficulty. The quality of film required is similar to that for urography.

4. Additional sedation is given as required.

5. When a pressure-injector is to be used it is made ready before the investigation starts.

TECHNIQUE

Equipment.—The basic equipment is an 18 G. arterial-cannula as used for arteriography, a flexible radio-opaque leader or guide-wire and the catheter. Details of this and related equipment are more fully discussed in the Appendix and in Part 2 of this chapter. A steel rule must be included on the sterile trolley for catheter measurement particularly when an examination without screening is performed, e.g. for placentography; it is also necessary when non-opaque catheters are employed. It should always be available in case the screening fails.

Contrast.—The volumes employed are similar to those used in direct puncture lumbar aortogram, p. 212.

Procedure.—1. An assistant scrubs up etc. and prepares the groin as before femoral puncture. A waterproof paper sheet is laid across the thighs and sterile towels over the abdomen and the legs, including the feet. The operator masks, dons a lead apron, scrubs up and puts on a sterile gown and gloves. When an image intensifier is being used he can proceed straight to arterial puncture and catheterisation; otherwise the procedure begins with measuring the catheter.

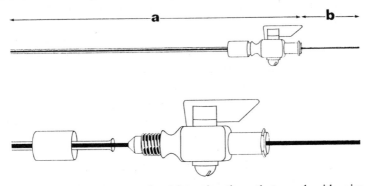

Fig. 37.—Catheter aortography. Measuring the catheter and guide-wire. The upper figure shows the two measurements which are made, *a*. the catheter length, *b*. the excess guide-wire when the guide-wire has been adjusted *so that its tip lies at the tip of the catheter*. The 'excess guide-wire' should exceed 15 cm. The lower figure is an 'exploded' view of a tap-adaptor, catheter with flange and the leader or guide-wire.

Measuring the catheters.—If necessary the catheter tip is first drawn out (see p. 237). Then, with the folding steel rule, two measurements are made:—

(*a*) The length of the catheter to its tap/adaptor (Fig. 37).
(*b*) The length of excess leader protruding from the tap end when the leader has been adjusted so that its other end lies at the catheter tip (Fig. 37). The leader is threaded into the catheter from the tip backwards so as to confirm that it will pass out smoothly through the tap. The 'excess leader' should be at least 15 cm. and is usually 20–30 cm. The leader should be worked through the catheter in both directions when using Kifa tubing since on occasions the leader cannot be withdrawn through the catheter when it has been passed into the artery. If the 'fit' is too tight immersion in hot water and further working of the leader will usually produce sufficient loosening.

These two measurements are written down. During all these manipulations care must be taken that free ends of leader and catheter do not swing against unsterile objects.

3. The catheter and leader are now separated, coiled loosely and placed

on the towels over the patient's legs. Towel clips are used to hold them down and prevent them from springing outward; one end of the leader should be near the groin so that it is 'to hand' as soon as the puncture is made. *When the metal guide-wire is used this must be the flexible end.*

4. The femoral artery is punctured wherever it is most easily felt; in palpating one should try to assess the line of the artery. A skin wheal of local anaesthetic is raised 2–3 cm. below the point chosen for puncture of the artery and deep infiltration is carried out around the artery as described in Chapter 27. Then the skin at the wheal is pierced with a small scalpel. The tip of the catheter is tested in the skin incision to make sure it will pass in easily.

5. The arterial cannula is taken up and checked to make sure that it is correctly assembled. A pad may be placed just below the site of puncture to absorb the blood that spurts from the cannula.

6. The arterial cannula is then thrust down through the artery, directing it along the line of the artery above the point of puncture. The stilet is then removed leaving the cannula in the tissues. It is slowly withdrawn until, often with a slight snapping sensation, the tip enters the arterial lumen and blood *spurts* from the hub. The importance of waiting for true arterial spurting must be stressed; a damped, pulsating flow may be obtained when the tip is only partly in the lumen.

The butt of the cannula is now gently depressed so as to point the tip more directly along the line of the artery. Blood should then spurt out in pulsating, vigorous gushes and the leader is taken up (*flexible end*, if a guide-wire) and passed 10–15 cm. into the artery. It is important to be certain that the flow is free and vigorous before introducing the leader and *the leader must slide freely into the artery with no sense of resistance. It must never be forced in.* The routine to be followed if there is any resistance is described below.

7. With the leader satisfactory in the artery the cannula may be withdrawn, avoiding any pull on the leader as it slides over it. When the cannula comes out blood may ooze from the puncture and pressure with a gauze swab is applied. The blood left on the leader by the cannula is wiped off. It has recently (Kay and Wilkins, 1969) been pointed out that wiping the serrated surface of a wound metal guide with a gauze swab leaves cotton fibres in the interstices of the wire. For this reason a non-adhering dressing soaked in saline or dextrose is used *with the plastic surface onto the guide.* (Sister F. I. Cole—personal communication). Alternatively a Teflon coated guide may be employed.[1] The fluid used for soaking the swabs must not be used for injection into the patient.

8. Now the catheter is threaded over the leader. *The leader must reach and protrude from the tap end before the catheter is pushed through the skin.* Sometimes the leader fails to pass through the tap/adaptor and the

[1] With Teflon coated guides a needle should never be employed for arterial puncture. Even the use of a cannula carries a slight risk of damaging the Teflon coating. A disposable needle-cannula is available with a plastic cannula (Cook Inc. Cat. no. 501–300) but is relatively expensive.

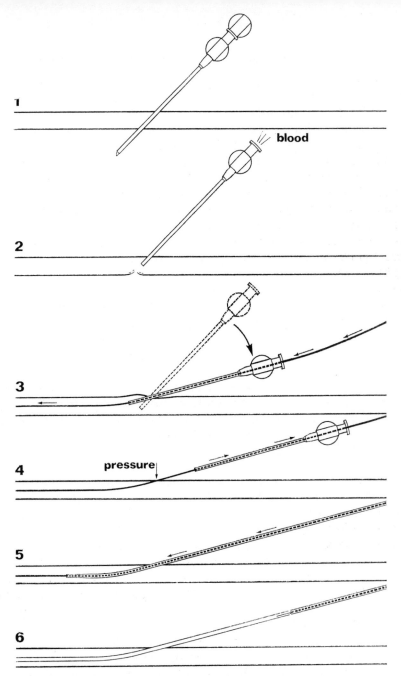

blood

pressure

FIG. 38.—Seldinger catheterisation. The sequence is:—1. The artery is punctured, usually by transfixion. 2. The stilet is removed and the cannula gradually withdrawn into the artery. 3. The butt of the cannula is lowered and the guide-wire passed carefully into the artery. 4. The cannula is withdrawn while pressure is exerted on the puncture site to stop bleeding and also to prevent accidental withdrawal of the guide-wire. The latter is wiped clean of blood. 5. The catheter is passed over the guide-wire and, after checking that the wire is emerging from its other end, it is passed into the artery. 6. After passing the catheter up the artery under screen control, the guide-wire is withdrawn.

latter is then unscrewed and the leader threaded through the components separately. Care should be taken to advance the catheter over the wire and not to pull the wire outwards.

Usually the catheter passes easily through the tissues and the artery wall. If there is difficulty it may be overcome by advancing the catheter with a rotary motion, gripping the catheter near the skin surface so that the catheter and leader advance together. Repeated gentle attempts seldom fail; should they do so the catheter tip should be inspected as it may have become torn or buckled and require drawing out afresh. Difficulty usually arises when using a catheter with a relatively thin wall which has been drawn out over a narrower guide-wire, e.g. O.P.P. 240 over a P.E. 160 (0·035″) guide-wire. It is then helpful to have prepared short (10–15 cm.) lengths of a tougher tubing, e.g. Kifa green, drawn out over the guide-wire. These can be used as dilators to stretch the puncture in the artery and so allow the thinner walled catheter to pass into the lumen without difficulty.[1]

9. Once the catheter is in the artery blood will begin to run into it around the leader unless there are no side holes at the tip. The catheter may now be advanced and the leader withdrawn alternatively until the leader tip is 5–10 cm. beyond the catheter tip and the catheter is advanced, *under screen control* to the chosen level. Between shifts of position the puncture site is pressed upon with a swab. This is seldom necessary for more than a few minutes but in some patients, particularly hypertensives or those on anti-coagulants, it may be necessary throughout the examination. When narrow catheters are used or when pressure has to be firm the leader is best left in the catheter to prevent kinking. When fluoroscopy is not available measurement of the catheter outside the patient enables one to adjust the tip fairly accurately since the total length of the catheter is already known. The aortic bifurcation lies at about 20–24 cm. from the point of puncture, the position for showing the renal arteries at about 32–36 cm. (according to the size of the patient) and the ascending aorta about 70–80 cm. When the position is thought to be correct the catheter is fastened to the thigh with sterile autoclave tape about 3–5 cm. from the puncture. This prevents accidental shift of the catheter. (Should the segment under the tape later need to be advanced into the artery it is cleaned with Hibitane/spirit and washed with sterile saline.

10. When the catheter is sited in approximately the correct position the leader is withdrawn; blood should then flow freely from the catheter and this is allowed to continue briefly to wash out any clot which might have formed around the leader. A syringe of contrast (Conray 280 or Hypaque 45) is attached, the catheter position is finally adjusted[2] and a brief injection is watched to make sure the catheter lies free in the lumen and is not engaged in a branch vessel. Contrast should be swept away immediately from the catheter tip except in aneurysms or in peripheral arteries which

[1] Teflon vessel dilators are now available, e.g. 7 F.G. dilator U.S.C.I. Cat. no. 7050 (G.U.) or Cook Inc. Cat. no. JCD–7–35 (Kimal Scientific Products, Ltd., 18 Pield Heath Road, Hillingdon, Middlesex).

[2] It is assumed that opaque catheter material is used for all investigations.

are the site of arteriectasia. When fluoroscopy is not available a film should be exposed during hand injection of 6–8 ml. of contrast. It may again be noted that in any department undertaking regular catheter work, fluoroscopic facilities are essential if catheters are to be manipulated with speed, safety and certainty. *A test injection must be watched or filmed before pressure injection* so as to prevent accidental injection into a branch of the aorta.

On rare occasions blood does not flow back when the leader is removed. If the catheter is truly within the lumen of the aorta this may be:—

(*a*) Because clots have formed in the catheter.
(*b*) Because the catheter tip is in a branch artery, or impacted against the aortic wall, possibly at an atheromatous plaque, or is lying subintimally.

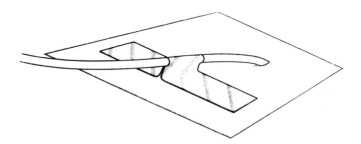

FIG. 39.—The use of sterile autoclave tape to fasten the catheter before pressure injection. The tape is prepared by rolling a short length on to a test tube and autoclaving it.

When this happens the catheter is withdrawn below the main visceral arteries and if there is still no flow an attempt is made to suck clot back into the syringe. This is not always possible and a small injection of dextrose may be needed to clear the catheter.[1] Should there still be no flow one must assume that the catheter is not in the aorta and it should be withdrawn slowly. Injection of contrast is *never* justified unless there is a good flow back. *When there has been any difficulty of this sort, even if a good flow is ultimately obtained, fluoroscopy or a test film during injection of 6 ml. of 45% Hypaque is mandatory to ensure that the tip lies in the aortic lumen.*

Once a satisfactory position has been achieved the catheter is strapped to the skin with sterile autoclave tape (Fig. 39).

11. Another point of importance must be stressed at this juncture because it arises in relation to the test injection of contrast; it applies, however, throughout any investigation in which a catheter is used. Whenever a syringe or flushing system is attached it is extremely easy to trap an air

---

[1] Although in theory this seems an unsound manoeuvre, in practice we have never found it to cause sequelae during or subsequent to an examination.

bubble which will then be injected. *Each time a syringe is attached, the air bubble should be drawn back into the syringe.* Similarly when the dextrose drip is connected the tap on the catheter should be opened to allow fluid or blood to drip from the catheter as the flushing system is inserted with dextrose dripping from it.

The practice of infusing dextrose or similar solution has been recommended in Chapter 27 as tending to reduce the risk of renal damage by contrast medium. It is now—i.e. after the test injection—a suitable time to carry out such instillation through the catheter.

12. When a test film is employed the catheter position is noted and if necessary altered. For demonstration of the kidneys the tip should lie between the upper poles of the kidneys, usually at about D.12/L.1. Contrast is now drawn up for the main injection, the centring of the film is checked, any 'coning down' needed is carried out and an assistant prepares the film changer. The patient is roused and told that he will shortly feel a warmth in his abdomen and down his legs. He is warned that he must hold his breath and keep still and that just before the injection is due he may be made to hyperventilate.

The syringe is connected to the catheter, carefully checking that air-bubbles are not included and that the tap is open and all connections are secure. When the Talley injection pump is used the syringe is placed in the cradle and the assembly is withdrawn until the catheter is as straight as possible. This is intended to prevent 'whipping' of the catheter during the pressure injection. The routine thereafter will vary with the nature of the examination and of the equipment used for filming. If the patient is conscious he is told to breathe in and out several times before being told to stop breathing; otherwise the anaesthetist is asked to arrest respiration for a suitable period. With hand injection the operator calls for the first exposure when about 5 or 10 ml. of contrast are left in the syringe. When only a limited number of films can be taken the next two should be exposed as quickly as possible and a fourth after a 4–5 second pause. With a film changer and pressure injection the routine is simpler. A satisfactory series on the changer is 2 films a second for 3 seconds followed by one film every 2 seconds up till 9 seconds. A 'count-down' technique is used and injection is commenced on the call 'one' while the series is initiated at zero.

After the last film is exposed the patient is told 'Breathe normally'. The injection syringe is disconnected and dextrose flushed through the catheter. The films are then awaited and repeated if necessary.

Special points

Injection.—Disposable syringes are satisfactory for all types of injection made by hand, especially when small volumes are required in selective arteriography and peripheral limb arteriography. For free injection into the aorta, especially when the narrower radio-opaque catheters are used, a pressure injector is essential and this also applies to coeliac and superior mesenteric arteriography. Details of the use of the Talley pump are discussed in part IV of this chapter.

Pelvic arteriography.—The catheter is passed up the iliac artery for about 24 cm. i.e. to just above the bifurcation of the aorta. When it is desired to show only the vessels of the pelvis the femoral arteries are occluded by pneumatic cuffs just before the contrast is injected, the cuffs being inflated until the foot-pulses disappear.

Renal arteriography.—This is one of the commonest indications for abdominal aortography. The catheter position has already been discussed.

For simple examination of the kidneys, e.g. for a renal mass, it is reasonable to cone down to the kidneys for the actual 'run', after the catheter position has been checked on the TV monitor. But where a test film is required it should include the whole abdomen as coning might 'cut off' the tip of the catheter. With a non-functioning kidney whose position is uncertain the pelvis should be included against the possibility of an ectopic kidney, and we normally include a wide field in a survey examination for hypertension since an unusually situated phaeochromocytoma will then be shown.

Abdominal compression.—A Bucky band with pads over the epigastrium improves definition of the kidneys, but is inadvisable in suspected renal tumours.

Oblique views.—These may be the only means of clearing the superior mesenteric vessels from the renal arteries or of clearing overlying gas from a particular area of kidney. 20° is usually enough. A difficulty arises in choosing which way to incline the patient for an oblique film since the orifices of the renal arteries may lie slightly anteriorly or posteriorly on the aorta. One cannot therefore say that elevation of the left side will necessarily show the orifice of the left renal artery better or vice versa. For this reason we normally choose whichever oblique position seems most likely to move overlying vessels clear of the renal arteries. Centring should be adjusted to allow for the obliquity. The simplest method may be to place the catheter at renal artery level under screen control and then take an 'undercouch' film in each oblique projection during hand injection of 20 ml. of Conray 280 or Hypaque 45.

Difficulties of the Seldinger method

1. *Failure to pass the leader into the artery.*—This springs from faulty positioning of the cannula. Normally the leader slides easily into and up the artery but various difficulties can occur.

A. *The leader may stop at the tip of the cannula.*

(i) Because the cannula tip is still partly in the opposite wall of the artery.
(ii) Because the angle of the cannula is too steep so that the leader impinges on the opposite wall or even enters the hole left by the puncture.

When this happens the leader is pulled out. The cannula is withdrawn a little and the butt further depressed. Provided the flowback of blood is good the leader is advanced again into the cannula. Continued difficulty in spite of a good flow at this stage should make one consider the possibility that the cannula is angled sideways to the artery and it may be swung first to one side and then the other, probing gently with the leader with each shift in the position.

Should these manoeuvres fail and the flow become weak it is better to pull the cannula slowly out and if the flow does not improve, attempt a fresh puncture. Sometimes bleeding may make it necessary to exert pressure for 5–10 minutes on the puncture site before re-puncture is undertaken.

B. *The leader may appear to pass into the artery but after a few centimetres is held up.* On rare occasions this is because the leader has passed subintimally but in most cases it is because the leader is caught at a bend in a tortuous iliac artery.[1] A catheter may then be slid carefully into the artery and the leader withdrawn; blood should flow back into the catheter confirming that it is placed in the lumen of the artery and one may then attempt to negotiate the kink. When a good flow back cannot be established the catheter and leader must be pulled out and preparations made for a fresh puncture.

2. *Difficulty with tortuous vessels.*—The first stage when the leader meets a tortuous iliac vessel is to probe gently with the slowly rotating leader watching the tip on the screen. The P.E. 160 (0·035″) leader which is normally used is fairly flexible and with gentleness and care will often pass through quite marked curves. If this fails the catheter is inserted as far as possible and the leader withdrawn 5–10 cm. from its tip. The catheter is used to probe the bends, watching its progress on the screen. It may help to remove the leader and inject a little contrast to show the direction of the tortuous vessel and to demonstrate any strictures or other abnormalities which may preclude successful passage of a catheter. If the catheter is supple or has a curve on its tip it will often pass up into the aorta but if not, a 'J'-shaped leader is used. This is a specially made leader whose curve is formed into a tight but flexible bend so that it is well suited to negotiating even quite severe curves. The catheter is drawn back a little from the obstruction and the J-guide is advanced 5 cm. beyond its tip. This assembly is then slowly advanced, rotating it until the kinks have been passed. An alternative, when a J-guide is not available, is to advance an OPP 160 catheter with a sharp bend at the tip, to about 1–2 cm. beyond the tip of the leader. The thin catheter is very supple and frequently bends over to form a type of J-guide. Should these manoeuvres fail the catheter and leader are removed. If at any stage pain is experienced as the catheter is advanced great caution should be exercised and if it persists one should pull the

---

[1] Very occasionally the leader passes into the deep external circumflex artery and becomes impacted. This can be recognised on fluoroscopy by its lateral position. It is carefully withdrawn into the femoral artery and advanced again, rotating it so as to direct its top more medially.

catheter out, press on the puncture site and prepare to puncture the artery on the opposite side.

Failure to catheterise on both sides is uncommon in the presence of good pulses. Should it occur one must decide whether to proceed to translumbar aortography[1] or axillary artery catheterisation. This decision will be affected by many considerations—the condition of the patient, the nature of the diagnostic problem etc.—so that no rules can be fixed. It is advisable not to proceed with direct translumber aortography if the blood pressure is very high since blood loss from an aortic puncture can be considerable in such cases and blood pressure control should be attempted for a period before translumbar aortography is undertaken.

3. *Difficulty with the leader.*—The danger of cutting the surface of a plastic leader has already been mentioned. Very occasionally the metal in a wire leader fails at the junction between stiff and flexible parts. The flexible tip may break off and form an embolus, necessitating removal— this is very rare. We have however seen an irreversible angulation at the same site so that the leader could not be removed through the needle or a catheter. The only safe method of removal is carefully to pull out the leader and catheter or needle *as one unit*—not allowing any movement to take place between the two.

Minor roughening of plastic coated catheters may occur after manipulation. Such leaders should be discarded.

4. *Kinking or rupture of the catheter.*—This is not normally a problem with Kifa or Teflon catheters but manipulation can kink or ruck a polythene catheter with local weakening of its wall particularly if the narrower (OPP 160, OPP 205) catheters are manipulated without a leader inside them. Another cause of kinking which can affect thick-walled catheters is when the wall opposite a side-hole is damaged by too forceful insertion of the punch used for making side-holes. When this happens the leader is advanced into the catheter and the catheter withdrawn and replaced. The new catheter should be measured before introduction if screening is not to be used.

Catheter rupture is only likely to occur when using a pressure injector. The common site is just beyond the tap adaptor. Unless rupture occurs very early during the injection it is best to expose as usual, for the films will often be quite satisfactory. The leader may then be re-introduced, and the catheter withdrawn and replaced (but see footnote 2 on p. 248).

5. *Subcutaneous coiling of the catheter.*—This can occur with Kifa catheters after prolonged manipulation and is commoner in fat patients or with haematoma formation. The operator finds that although he advances

[1] Direct retrograde femoral aortography (Waldhausen and Klatte, 1962) is another alternative. A PE 205 needle or cannula is inserted into a femoral artery. The thigh vessels are occluded with pneumatic cuffs, the patient performs the Valsalva manoeuvre and 50 ml. of 45% Hypaque mixed with 30 ml. saline are injected in two seconds. The first film is exposed when 10–15 ml. are left in the syringe. We have no personal experience of this method but a similar approach has been used with success when a pressure injection has been made through a P.E. 240 catheter positioned in the iliac artery but unable to reach the aorta. Before trying this method it is advisable to watch the effect of a hand injection on the TV monitor.

or withdraws the catheter through the skin no movement of its tip appears on the screen. The catheter curls into loops under the skin, having become softened by body warmth and continued manipulation. It may be possible to carry out successful manipulation after pulling out the loops and inserting a metal guide-wire. If this is not successful it is advisable to change the catheter.

Removal of the catheter.—When the examination is concluded the catheter is withdrawn to the external iliac artery and the foot-pulses are again palpated while pressure at the site of puncture is released. If the pulses are normal the catheter is pulled out. Should they show signs of impairment it is advisable to follow the routine set out on p. 202. Before doing this the catheter is withdrawn until only 5–10 cm. lie in the artery.

The catheter may now be removed and manual pressure exerted on the puncture site. The pressure should cover both the skin wound and the estimated site of arterial puncture. As a routine we invariably compress for at least 10 minutes after using OPP 205 or OPP 240 but much longer periods may be needed following prolonged manipulation of the various Kifa catheters or in hypertensive subjects. The use of a Bucky band and fluid-cushion for more effective compression is described on p. 200. When pressure is released the site is watched and gently palpated for a further minute to make sure no haematoma is forming. Injection of hyalase 1,000 units in 10 ml. water for injection may be used when there is a significant haematoma.

After-Care.—As for femoral arteriography.

References

KAY, J. M., and WILKINS, R. A. (1969). *Clin. Radiol.* **20**, 410.
SELDINGER, S. I. (1953). *Acta Radiol.*, **39**, 368.
SUTTON, D. (1962). *Arteriography*. (E. & S. Livingstone, Edinburgh).
WALDHAUSEN, J. A., and KLATTE, E. C. (1962). *N. Engl. J. Med.*, **267**, 480.

Further Reading

BOIJSEN, E., and FEINSTEIN, G. L. (1961). Arteriographic Catheterisation Techniques. *Amer. J. Roentgenol.*, **85**, 1037.
Thoracic Aortography: A symposium. (1960). *Br. J. Radiol.*, **33**, 531.
    I. Thoracic Aortography in Adults—Technical Aspects: D. McC. GREGG.
   II. Thoracic Aortography in Adults—Clinical Aspects: D. SUTTON.
  III. Thoracic Aortography in Infants and Young Children: R. E. STEINER.
  IV. Thoracic Aortography and Cine-radiography of the Aortic Valve; K. E. JEFFERSON.
SAMUEL, E. (1962). Percutaneous Brachial Artery Catheterisation. *Br. J. Radiol.*, **35**, 468.

APPENDIX

Ward instructions

As for direct puncture aortogram—the pubic shave must be included.

After-care

As for femoral arteriogram.

Trolley Setting for Seldinger Catheterisation

Basic setting plus:

Seldinger arterial needle P.E. 160.
or Thin wall 18 G. arterial cannula.
Guide-wire (leader) size P.E. 160 (·035″ D.) at least 20 cms. longer than the catheter.
Catheter tapered to guide-wire P.E. 160 and flanged.
Catheter adaptor.
50 cm. length of guide-wire P.E. 160 (or fuse wire 21 G. (30 amp)) for forming bend at tip of catheter.
Catheter punch if side holes are required.
Artery forceps if polythene catheters are tapered at time of investigation.
Folding metal rule.
Non-adherent dressing 4″ × 4″.
Complete unit × 500 ml. Steriflex lotion for flushing catheters (used with Fenwall pressure bag).
2-way tap adaptor.
Roll of sterilised autoclave tape.

Also required

Talley pump or other injection device. When the Talley pump is used compressed air is the safest gas to employ. A spare cylinder should be available.
Source of boiling water, for forming bend on catheter.
Fenwall pressure bag (Baxter laboratories).
Stand for suspending Fenwall bag ('drip' stand).

N.B.—Heparinised dextrose or saline is made up with 10,000 i.u. of heparin to the litre. Injection of more than a litre is to be avoided as generalised effects may be produced.

It is advisable to have available a supply of sterile, tapered and flanged catheters of assorted types and lengths; similarly a variety of guide-wires should be available.

'J' Guide-Wires

Small/large curved safety 'J' wire guide.

Makers: Cook Incorporated.

Suppliers: Polystan (G.B.) Ltd.,   *or* Kimal Scientific Products, Ltd.,
Wilbury Way,   18 Pield Heath Road,
Hitchin,   Hillingdon,
Herts.   Middlesex.

SELDINGER CATHETERISATION

Part II—The preparation and care of catheters

Catheter materials.—In spite of the advent of newer materials such as Teflon (tetrafluorethylene—PTFE), polythene is still the commonest catheter tubing for vascular work. The range of sizes and types of tubing may seem confusing (Table VI—General Appendix) but, in fact, the commoner examinations can be performed with a relatively restricted number of types of catheter. In this section we discuss the preparation of polythene catheters in the belief that even when they normally use preformed catheters[1] radiologists should have some knowledge of the preparation of catheters:—

1. Because they may at times need to train assistants who will have to prepare their own catheters.
2. Because the ability to produce a suitable shape in a catheter may be of critical importance in achieving successful selective catheterisation.

The radiologist who can carry out the procedures set out in this chapter will be able to meet almost any demand likely to arise in routine vascular work. He will also be better able to assess and use the more expensive preformed catheters for specialised use (see the manufacturers listed in the Appendix to this chapter).

We propose to mention only one size of guide-wire—the size commonly referred to as P.E. 160 because of the size of polythene tubing to which it initially corresponded. It is now accepted that any size of tubing can be 'drawn out' to fit snugly over this size of wire. This has several advantages: i. the guide-wire is slender and relatively atraumatic, ii. the needle used is narrower so that the puncture made in the opposite wall during transfixion of the artery is correspondingly smaller.

Only two groups of catheter material will be discussed—the opaque

---

[1] *Pre-formed Catheters*
    Makers: U.S.C.I.—Kifa.
    Suppliers: Chas. Thackray,
        P.O. Box 171,
        Park Street,
        Leeds 1.

    Makers: Cook Inc.,
    Suppliers: Kimal Scientific Products Ltd.,
        18 Pield Heath Road,
        Hillingdon,
        Middlesex.

polythene produced by Portex[1] and the Kifa range of polythene. The opaque Portex is a softer material, less densely opaque and less able to maintain a curve into which it has been formed. It does not withstand high injection pressures so well as the Kifa tubing but its wider bore compensates for this. Its qualities make it suitable for abdominal aortography and catheterisation of peripheral arteries or veins. The green and red Kifa tubing (Table VI—General Appendix) are generally used for selective work, particularly in the abdominal aorta. They are of equal lumen but the green has a thicker wall which make it easier to manipulate at a distance. The yellow and grey Kifa tubing have the same external diameter but the grey has a wider bore. It is generally employed for arch aortography where a large volume of contrast must be delivered in a short space of time. The yellow tubing is now little used but its thick wall enables it to hold complicated curves and may make it easier to control the tip at a distance.

Preparation of polythene catheters.—The preparation of catheters may be carried out at the time of the examination but we have found it more convenient and less time-consuming to prepare catheters in standard lengths in advance and to store them after sterilisation, preferably by ethylene oxide (see p. 31). The various procedures in preparation are:—

1. drawing out the tip.
2. punching side holes.
3. flanging the catheter.
4. shaping the catheter.

Of these procedures the first three are most suitably performed well before the examination. Shaping the catheter may be done at the same time, but is a very simple process and is best carried out at the time of the examination when the size of the patient and the nature of the diagnostic problem will be known. Instructions for the preparation of Kifa opaque catheters are given with each coil of catheter tubing (G.U.; S.X.). It should be pointed out that punches for making side holes, a flanging tool and suitable emery paper (G.U.; S.X.) should be purchased when the first coils are bought.

Drawing out the catheter.—In order to make it easier and less traumatic to insert the catheter through the arterial wall the tubing is drawn out to a tapering end which fits snugly over the leader. To draw out the opaque Portex tubing a wire leader is passed into a length of tubing which has been cut about 10 cm. longer than the final length desired for the catheter. The stiff end of the leader is adjusted to lie about 6 cm. short of the tip of the catheter. The catheter end is gripped with an artery forceps held in any convenient fixed point while the catheter is tightly gripped on to the leader near its tip with both hands (Fig. 40). A firm pull is given, drawing out the catheter tubing and tapering it. Using this method the trans-

---

[1] Remarks made about this type of opaque polythene can in general be taken to apply to non-opaque polythene of the P.E. (Intramedic) or PP (Portex) ranges.

itional area from full width to maximum narrowing is fairly short. The leader is pulled back and the catheter is cut across towards the narrow end of the tapered zone. The tip so produced should be firm with a short neat taper and it should fit closely to the leader. It is possible to draw out OPP 240 on to a P.E. 160 leader by this technique, but it is much more difficult even if the tubing is first warmed, although it may be achieved

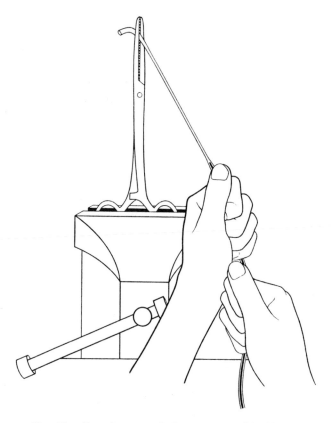

FIG. 40.—Drawing out polythene over a guide-wire.

with practice. To make this operation simpler we have developed a former (M.D.) on to which the wide polythene can be easily pulled out and which will produce a satisfactory tapering 'snout'.

To draw out the Kifa tube requires heat. A methylated spirit lamp or gas burner are adjusted to a low flame. A length of P.E. 160 guide-wire is inserted into the end of the tubing which is then held between the two hands about 2–3 cm. over the flame, rotating it *evenly* until it softens. It is then removed from the flame and drawn apart to produce a tapered area; it should be held under cold water in this position until cool. The wire is then withdrawn and the narrower area cut across at a suitable point with a sharp blade. The guide-wire should be worked back and forth in the tapered end to make sure that the 'fit' is not too tight. It is easy to overheat the tubing causing the catheter to break and some practice is required before a taper can be easily produced. The thicker tubing, i.e. grey and yellow, should be heated over a wider area so as to produce a smooth taper. The most important factor in producing a smooth, even taper is to apply the heat uniformly over the whole circumference before drawing the tubing apart. The catheter tip is inspected and any irregularities are removed with a cosmetic emery board, followed by no. 4 Kifa emery paper.

Punching side holes.—Side holes are desirable when the catheter tip has been tapered to any significant extent since they allow a greater area for escape of contrast from the end of the catheter. Further, they may help to reduce recoil, particularly if they can be suitably angled. Side-holes are only practicable in the thicker catheters whose walls will not be unduly weakened by them. They are therefore made in the Kifa catheters and the OPP 205, 240 and 270. When punched in the OPP 205 or 240 they should be made away from the tapered area since holes here will weaken the tip. Punches should be kept sharp so that holes are made cleanly and without undue effort. Great care must always be taken to ensure that the small plugs of polythene are not left in the lumen of the catheter, nor projecting to roughen the outer wall. It is most important *never* to damage the opposite wall nor to puncture both walls at one site so weakening the catheter to a dangerous extent. The punches used are those intended for preparation of Kifa catheters (G.U., S.X.).

Flanging the catheters.—The tap-adaptors used with these types of catheter require a flange on the end of the catheter to give a firm grip (Fig. 37). For clear polythene it is achieved simply by bringing the cut end of the tubing gently towards a flame. The heat produces an outward splaying of the end of the tubing. A spirit lamp or candle is a convenient source of flame and is used in a corner of a room free from draughts. Some care is needed to avoid overheating and melting the catheter. The underside of the flame is the steadiest part; the catheter end (the tubing having been

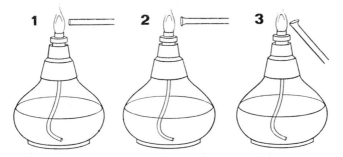

FIG. 41.—Flanging polythene catheters. 1. The polythene is brought gently towards the flame. 2. An even flange usually results. 3 If the flange is 'tilted' it may be possible to straighten it by bringing the projecting edge nearer to the flame.

drawn out and cut to the chosen length) is brought up slowly at a right angle to its surface. With a little practice it becomes easy to obtain an even flange (Fig. 41). Sometimes the flange is 'tilted'; when this occurs the most prominent side of the flange is put nearer to the heat and the flange will usually straighten. The opaque polythene, particularly the Kifa tubing, does not produce a satisfactory flange with this method alone. A small flange is made in this way but is then enlarged by inserting a conical flanging tool (G.U.; S.X.) which has been heated in a flame. The tool is pressed into the tubing to widen the flange and then cooled in cold water. It can then be removed and the flange inspected to make sure that it is satisfactory. All flanges should be fitted into a tap-adaptor to make sure they are not too wide, and until one is practised it is advisable to test the whole catheter-tap-adaptor assembly by a forced injection through it.

Shaping the catheters.—This procedure is most frequently used in preparing catheters for selective arteriography but catheters must also be shaped if they are to pass round the aortic arch and a curve is often necessary to help the catheter through tortuous iliac vessels or round the sharp bend where the subclavian artery arches over the first rib.

The simplest way to produce a curve is to insert a 50 cm. length of P.E. 160 guide-wire (a portion of discarded wire is suitable) about 30 cm. down the tapered tip of the tubing. If the free end of the guide is gripped against the stem of the catheter a loop results (see the illustrations on the Kifa packet). The radius and configuration of this loop can be adjusted to give a wide variety of shapes. The curved part is dipped into near-boiling water for 30 seconds and maintaining the curve, immersed in cold saline or dextrose on the trolley. It should then maintain the shape which has been set in it after withdrawal of the leader. If the *tip* of the catheter is put in hot water it becomes softened and the tension on the wire widens the mouth making it harder to insert in the artery. When a complex shape is required we have found it is most easily obtained by inserting malleable fuse wire (30 amp., gauge 21) into the catheter before shaping and immersing in near-boiling water in the usual way.

This shaping may be carried out at the outset of the examination under aseptic conditions or it may be left until a 'free' injection in the lumen of the aorta has shown its width and the shape of origin and number of the vessels to be catheterised.

Guide-wires (leaders).—Although the only gauge of wire to be discussed is the P.E. 160 (diameter 0·87 mm.; 0·035 inches) a number of varieties are available. A short (15 cm.) length guide with an attached needle-stop is used to pass the cannula portion of an arterial cannula well into an artery (see p. 199). For catheterisation of the aorta a 120 cm. guide is often adequate but a 150 cm. guide has certain advantages. For example, if there has been difficulty in passing the catheter through tortuous iliacs and one then wishes to change the catheter—say from an O.P.P. 240 used for a free injection to a green Kifa for selective renal arteriography—the longer leader will still have its tip in the abdominal aorta when the first catheter has been completely withdrawn. This saves the need for a potentially traumatic and difficult second attempt at traversing the iliacs.

The standard wire construction is a solid wire core with a spiral winding firmly surrounding it. A short section at the tip has no core and is therefore flexible. This is the weakest point of the guide and in certain makes a linking metal strip is included to prevent it becoming detached. Guides are also available with a moveable core so that the length of the flexible part

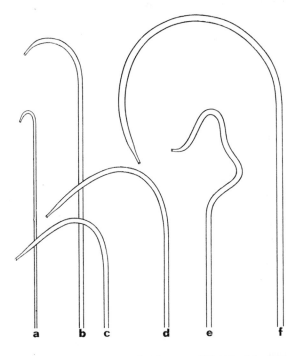

½ scale
(approx.)

Fig. 42.—Various types of catheter. *a*. PE 160 with curve suitable for entering a child's renal artery. *b*. suitable curve for catheterising branches of the adult aorta—green or red Kifa. *c*. and *d*. curves suitable for the right and left renal veins—green or grey Kifa. *e*. Mikaelsson catheter, mentioned in the text. *f*. Grey Kifa with a curve for entering the ascending aorta.

can be altered. The most useful of the specially shaped guide-wires is the J-guide whose use in circumventing kinked iliac arteries has already been mentioned. Other guides are available with pre-formed curves intended temporarily to increase the bend at a catheter tip: a simpler way of achieving this is to bend the 'stiff' end of a sterilised guide-wire to a suitable curve and advance it to within a few millimetres of the end of the catheter (Fig. 43).

Length of catheters.—The length of catheter required varies with the area to be examined and the route used. Normally for abdominal aortography we use a catheter length of 70 cm. and for aortic arch injection a length of 100–110 cm. Lengths in excess of what is needed are undesirable since the resistance of a tube of constant bore increases with increase in length.

Sterilisation and storage of catheters, connections and leaders.—The problems of catheter sterilisation have been discussed in Chapter 4. In the absence of facilities for gamma irradiation or ethylene oxide sterilisation, we suggest the use of chlorhexidine/spirit to immerse and to instil through the catheter. Leaders and arteriographic connections are sterilised by autoclaving.

SELDINGER CATHETERISATION

Part III

Notes on catheter manipulation and selective catheterisation

It is not possible to mention every situation in which selective catheterisation is undertaken but some general remarks may be of value. To begin with it should be pointed out that in many investigations, particularly of the renal vessels, a free injection of contrast and series of films is an essential preliminary. It shows the level, number and configuration of the branches of the aorta. One point of particular importance is the demonstration that an artery shows early branching. This warns the operator against inserting a selective catheter too far and so missing part of the distal circulation. The free injection series also shows the width of the aorta which is a guide to the width of the corresponding selective catheter. Even when a full series is not thought necessary it may be helpful to take a film on the under-couch tube during hand injection in the area of greatest interest.

Catheter shape.—For most arteries a simple curve (Fig. 42b) is satisfactory but the diameter of the curve should be increased when the aorta is wide and reduced when it is narrow. If the curve is too wide for the

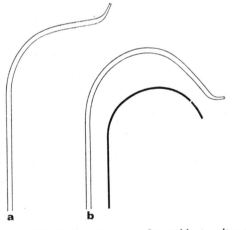

a   b

FIG. 43.—Illustrating the use of a guide to alter the curve of a catheter. *a.* shows the shape of a catheter designed to enter the left adrenal vein. *b.* shows how a curved guide will on insertion produce a curve suitable to direct the catheter into the left renal vein.

aorta the tip becomes bent right over and often will not engage in a branch (Fig. 44b). Even if it does engage it may pass too far into the vessel. A curve which is too narrow for the aorta, on the other hand, may make it impossible to direct a catheter tip against the wall of the aorta. One may be able to overcome either difficulty as outlined below but often it is better to prepare a fresh catheter more suitably shaped and then exchange catheters. When the curve is too tight the guide-wire can be re-inserted to straighten it out at a point above the artery to be entered. The guide is then removed leaving the catheter 'opened out' and with its tip pressed against the aortic wall. It can then be withdrawn slowly to the region of the chosen artery, rotating it to bring the tip to the appropriate site. A slight 'jigging' up and down movement may then cause it to enter the artery. Sometimes the reverse occurs and the catheter will not bend over into its usual curve (Fig. 45). It is then withdrawn to the lower abdominal

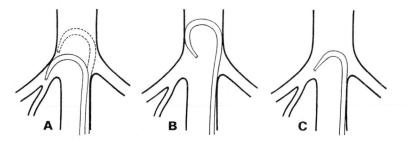

Fig. 44.—A. The correct size of curve makes catheterisation easy. B. Too wide a curve causes the tip to curl over. C. Too small a curve makes it impossible to engage in branch vessels.

aorta and advanced up one or other side wall in the hope that its tip will engage in a branch such as a renal vessel. When it does so the catheter is advanced further (if necessary with a guide-wire advanced close to the tip to stiffen it) so that the tip pulls out of the branch, leaving the catheter with the curve satisfactorily formed. A catheter whose curve is too narrow may be improved by bending an old (but sterilised) guide-wire to a better shape and advancing it to within a few mm. of the tip. Although both manoeuvres mentioned can be successful too much time should not be spent before exchanging for a better shaped catheter.

Other shapes of catheter and their suggested use are shown in Fig. 42. An alternative general purpose curve devised by Dr. Mikaelsson of Stockholm is also shown. Our experience with it is limited but its advantage is that it is more stable once engaged in an arterial ostium and that it is more adaptable to variations in width of the aorta.

Manipulating the catheter.—The catheter is advanced over the guide to the approximate region of the selected vessel. The guide is withdrawn; this usually allows the catheter to assume its chosen curve, with the tip directed downward and laterally. The flushing drip is now attached. If the catheter fails to take up the curve it is advanced into the thoracic aorta where the greater width may allow it to bend over. Alternatively an attempt may be made to engage the tip in any branch vessel, as described earlier, and so hold the tip while the rest of the catheter is advanced upwards.

Those who are beginning to learn selective catheterisation should not go further without inspecting the films taken during free injection. It is particularly important to make sure that all the branches feeding a particular vessel, e.g. kidney, have been recognised, otherwise selective

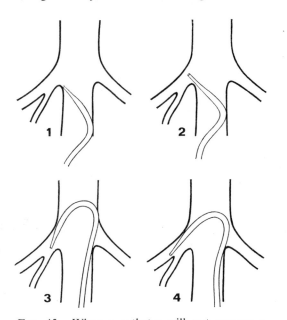

Fig. 45.—When a catheter will not assume a satisfactory curve it is—1. withdrawn into the lower aorta. 2. moved upward until the tip engages in a branch, e.g. a renal artery. 3. it is further advanced, if necessary stiffening with a guide-wire, until it curves over. 4. shows how too large a curve may cause the catheter to pass far down an artery, so that an early branch vessel may not be filled.

catheterisation of the most important vessel may not take place and it is even possible to regard an area of avascularity as pathological when it simply represents an area whose blood supply has not been demonstrated. The catheter tip is now brought to the region of the selected artery and manoeuvres to engage it in the vessel then take place. For example, in catheterisation of a renal artery the catheter is held so that its curve is seen in profile and with a 'jigging' motion is advanced and withdrawn along the side wall of the aorta. Entry into a branch of the aorta is indicated either because the tip suddenly moves outwards or because the tip is

I

held at a certain point while the rest of the catheter moves up and down. Outward movement is most obvious with laterally placed vessels and is not seen when the artery is directed anteriorly or posteriorly. Even renal arteries may come off the anterior or posterior quadrant of the aorta and if the catheter fails to engage at the expected level it is rotated a little anteriorly or posteriorly. As the rotary movement is imparted it is advisable to advance and withdraw the catheter a little; otherwise the tip tends to stick at one point and suddenly rotate more than intended. In all manipulations the catheter should be held close to the puncture site while the left hand gently presses to check any buckling. Only a small area need be watched on the fluoroscopic screen—it may take some time to enter a particular vessel and radiation should therefore be kept to a minimum by this means.

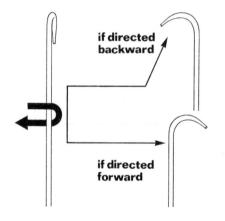

if directed
backward

if directed
forward

Fig. 46.—The direction of a catheter tip can be inferred from its movement on rotation. It is, for example, watched during clockwise rotation and the way the tip moves indicates whether it was pointing forward or backward. It can then be rotated in whichever direction is required.

Sometimes difficulty is encountered in directing the catheter tip towards the right side after the left has been entered or vice versa. This is particularly true if the aorta or iliacs are tortuous and it may, in fact, be necessary to fashion a fresh catheter with its curves adapted to the abnormal anatomy, often including extra curves along the shaft of the catheter. It is then exchanged, using a guide long enough to keep its tip above the tortuous area even when the catheter has been completely removed. Before this is done, however, it is worthwhile placing the catheter with its tip at the appropriate level and rotating it several times in each direction, advancing and withdrawing slightly as described earlier. The tip will sometimes spin round and enter the desired vessel making it unnecessary to change the catheter.

Beginners often find it difficult to tell whether a curved catheter is lying with its tip directed anteriorly or posteriorly. This is usually not difficult (Fig. 46). If, for example, the catheter is rotated in a clockwise direction the tip will turn to the right if it is directed posteriorly and to the left if anteriorly. Anticlockwise rotation has the opposite effect. With practice the assessment of position in this way becomes almost instinctive.

When it is thought that the tip is engaged in a branch vessel the flushing drip is stopped and disconnected and a syringe of contrast (Conray 280 or Hypaque 45) is attached (alternatively a 3-way tap may be used). As mentioned earlier, an air-bubble is almost always included and *should be drawn back into the syringe.* If one cannot withdraw an air-bubble, easily followed by some blood, it suggests that the catheter orifice is pressed against the wall of the aorta or impacted in a narrow branch, e.g. a lumbar artery. Contrast should not be injected until further manipulation has produced a flow-back of blood. Lumbar (or intercostal) artery injection is very painful and should be avoided in the conscious patient.

Catheterisation of anteriorly or posteriorly directed branches, e.g. coeliac axis or superior mesenteric arteries, follows similar principles but if it is not successful a lateral film during free injection should be obtained. This may show either a narrowing or occlusion of the vessel or may demonstrate an unusual anatomy—for example, the coeliac axis is some-times directed upwards rather than downwards and may require a differently shaped catheter.

Obtaining films after selective catheterisation.—For certain problems, particularly those arising in relation to the vessels to the gut, a series of films with the over-couch tube and pressure injection is likely to be necessary, e.g. for a hepatic arteriogram. With other problems, par-ticularly in relation to the kidneys, films taken with the under-couch tube are usually adequate. As soon as the correct positioning of the catheter in a renal artery has been confirmed by brief injection of a small amount of contrast, the diaphragms are adjusted to include the length of the renal vessel and the whole of the kidney outline at the position of the kidneys when the breath is held (or when the anaesthetist arrests respira-tion). The average kidney, supplied by a single renal artery, is adequately shown by brisk injection of 8–10 ml. of Conray 280 or Hypaque 45. One film is exposed towards the end of this injection. The film is then exchanged as rapidly as possible and a second film exposed usually 7–10 seconds later. The latter shows the nephrographic phase and will often demon-strate opacification of the renal vein. With multiple renal arteries pro-portionally smaller amounts are used. For all renal masses it is advisable to obtain films both in the AP and in *both* oblique projections. Only in this way will all contours of the mass and its vascular supply be properly shown. Occasionally in a case in which the mass is of obscure nature a lateral film may be necessary and this usually requires an 'over-couch' type of film.

In all cases the opposite renal artery should be injected so as to exclude an unnoticed lesion on this side. When a renal carcinoma has been shown a further series of films is taken after injection of 30 ml. or so of contrast so as to outline the renal vein and show any tumour extension. If the lesion appears avascular it should be punctured (see Chapter 9), preferably at the same session as the arteriogram.

SELDINGER CATHETERISATION

Part IV—The use of the Talley pump[1]

THE first feature of the Talley pump which must be understood is the relationship between operating pressure and size of syringe. The various syringes differ not in length but in diameter. The effect of this is that a much higher pressure is created in the small than the large syringes for a given pressure in the pneumatic ram. One should therefore always specify the size of syringe used when discussing injection pressures; a pressure which is suitable when using a 50 ml. syringe will usually rupture the catheter if applied to a 25 ml. syringe.

With any catheter of average length or greater it is found that during the turbulent flow which occurs with pressure injection, the pressure gradient along the catheter is comparable to that existing when injection is made into a catheter whose end has been blocked. Thus to determine the maximum safe pressure to avoid catheter rupture, measurements should be made with catheters whose end is occluded. Such figures, given in the table at the end of this section, represent safe working pressures for everyday use; they can be exceeded without necessarily producing catheter rupture but always with a risk of rupture. Should a catheter burst during injection, commonly at its proximal end, it is advisable to continue with the film series as sufficient contrast is usually found to have been injected.[2]

Because the syringes are made of metal it is not possible to see whether all bubbles have been expelled before injection. The syringes are constructed to permit complete removal of air after contrast has been drawn up. Any radiologist using the Talley system should first make sure that he understands the method of emptying air from it.

Care should be taken to ensure that all the components of a particular syringe are marked so that the matching plunger and barrel are correctly reassembled after sterilisation. To facilitate the passage of the plunger through the barrel a little sterile liquid paraffin is spread on the plunger with a swab; the plunger is then worked up and down the barrel until it moves easily: it is taken out to allow the contrast to be poured into the barrel. Contrast is easily poured from a vial; when the contrast is in an ampoule pouring is more difficult unless file cuts are made on *the shoulders*

---

[1] Supplied by Talley Surgical Instruments Ltd., 505 Liverpool Road, London N.7.
[2] When a catheter ruptures it can be immediately exchanged, using a guide-wire in the usual way. It may, however, be more convenient to cut across the catheter and insert a wide-bore needle, e.g. a drawing-up cannula. Patency can then be maintained until it is seen whether further films are required.

of the ampoule and its top is knocked off sharply. This provides a wide mouth through which the contrast flows readily. When the syringe is assembled and all air has been expelled the contrast is warmed approximately to blood heat by immersing the syringe in a steriliser containing boiling water.

When connecting the catheter to the syringe all steps should be taken to reduce any impedance to flow such as unnecessary bends in the catheter, tap-adaptors which are narrower in gauge than the catheters themselves or narrowing at the catheter tip without compensating side holes. In practice it is usually desirable to achieve injection in the shortest possible time and thus with the highest possible pressure.

INJECTION TIMES AND PRESSURES WITH THE TALLEY PUMP[1]

Catheter Size & Type	Diameter Int. mm.	Diameter Ext. mm.	Safe Operating Pressure 50 ml. syringe P.S.I.	Safe Operating Pressure 25 ml. syringe P.S.I.	Safe Operating Pressure 10 ml. syringe P.S.I.	Time for 30 ml. Contrast — Secs.
P.P. 100	0·86	1·52	80	32	16	8
(0)P.P. 160	1·14	1·57	35	14	7	5
N.I.H. 6	1·17	2·00	90	36	18	2·3
Kifa Red	1·2	2·20	80	32	16	2·3
Kifa Green	1·2	2·40	90	36	18	1·7
(0)P.P. 205	1·57	2·08	40	16	8	1·6
(0)P.P. 240	1·67	2·42	45	18	9	1·5
N.I.H. 7	1·47	2·33	90	36	18	1·3
Kifa Yellow	1·5	2·80	70	28	14	1·3
Kifa Grey	1·8	2·80	50	20	10	1·0
N.I.H. 8	1·73	2·67	90	36	18	1·0
P.P. 260	1·77	2·80	50	20	10	0·9
Teflon 8	1·73	2·67	90	36	18	0·9
N.I.H. 9	1·98	3·00	90	36	18	0·9

1. Injection times are for 85% Hypaque at 37° C and are similar for Urografin 76%, Triosil 70% and Conray 420.
2. Injection times are proportional to the volumes injected.
3. The length of polythene tubes for the above times is 100 cm. Halving the length decreases the injection time to about ¾ of the above times.
4. The pressures given are just below the static bursting pressures of the tubes and cannot be exceeded without some risk of catheter rupture. Halving the pressure increases the injection times by about 50%.

[1] This Table is reproduced by kind permission of Dr. John Dow and the Hospital Physics Department, Guys Hospital.

CHAPTER 30

LOWER LIMB PHLEBOGRAPHY

Oscar Craig, F.R.C.S.I., F.F.R.

MANY methods for lower limb phlebography have been described during the past few years. They have consisted of the intra-osseous route, retrograde injection of the femoral vein, ascending phlebography, supine and erect, and popliteal vein injection. It appears that no single method will give all the necessary information in each case, and the technique may have to be varied. When one single method is chosen, the following procedure has been found to be the most satisfactory in giving the maximum information for routine clinical practice. The technique is based on phlebographic work of Dow (1951a, 1951b), Gryspeerdt (1953), and others, with modifications developed over the past ten years. The method consists of an ascending injection, with compression of the superficial veins at the ankle and below the knee to force the contrast into the deep venous system. Many workers have found that the ankle tourniquet is the major source of unsatisfactory films, and adjustments to it may be difficult, resulting in either complete occlusion of the venous pathway or flooding of all the leg veins (Halliday, 1968). A further disadvantage was that pelvic phlebograms were difficult to obtain by contrast injections at the ankle. By the use of narrow inflatable cuffs compression at the ankle and knee can be varied. It is thus easy to control the venous filling and ensure selective deep vein visualisation. This filling can be observed and checked by fluoroscopy. It is also possible to control the flow of contrast sufficiently to obtain satisfactory visualisation of the external and common iliac veins (Craig and Snell, 1966).

Lower limb phlebography is not an essential precursor to operation in the majority of cases, but it is of value:—

1. In determining the site and extent of incompetent perforating channels in patients with varicose veins, information which is especially helpful in cases of recurrent leg ulceration following previous operations.

2. In determining the presence and competence of the deep venous valves. This applies particularly to those cases which present with oedema and in whom no visible superficial varicosities are found.

3. In the investigation of cases of acute or recurrent pulmonary emboli where peripheral venous thrombosis is likely and venous ligation is considered necessary.

4. In the investigation of cases of ulceration or oedema following deep vein thrombosis.

5. In localising accessory saphenous veins, especially in cases of recurrent varicosities.

6. In examining the iliac veins for thrombosis in cases of oedema, ulceration, recurrent varices and pubic varices, especially where these have arisen following a previous vein operation.

Contra-indications

(*a*) Known sensitivity to contrast media.

(*b*) Extensive recent thrombophlebitis.

(*c*) Special care should be taken in cases with a high prothrombin time following anticoagulant therapy, because of the risk of haematoma formation.

Preparation

1. Phlebography can be performed on out-patients. It is not necessary for the patient to be admitted overnight, but where pre-medication is used, time should be allowed for adequate recovery after the examination.

2. Food is withheld for six hours and fluid for four hours before the examination.

3. Anticoagulants are stopped temporarily if the clinician in charge of the case is satisfied that it is safe to do so. Phenindione (Dindevan) should be stopped for 36–48 hours and heparin for 6–8 hours prior to the examination. The prothrombin and clotting times should be performed as a check in these cases. If it is inadvisable to stop the anticoagulants and the examination is still required then a careful attempt at percutaneous venepuncture can be performed. If there is failure to perform a 'clean' puncture after the initial attempt, it is preferable immediately to perform a 'cut-down' and insert a cannula. It is usually easy to control the bleeding from the small incision necessary and the vessel is ligated on completion of the examination.

4. In very apprehensive patients sedation may be used, e.g. Nembutal 200 mg. one hour before the investigation; occasionally further sedation is required. In the vast majority of patients, if the examination is carefully performed sedation is not required. There should be little more discomfort than in performing an intravenous urogram.

5. The bladder should be emptied to ensure greater comfort for the patient during the examination.

Equipment

Some appliance for obstructing the superficial venous system above the ankle and below the knee should be available. In the past thin rubber tourniquets have been used, but as already noted, it is difficult by this means to control the degree of compression with any precision. It has been found more convenient to employ two thin pneumatic cuffs one above the ankle and the other below the knee. These are connected by rubber tubing to a clock-faced manometer and a hand bellows. Each cuff is controlled by a separate stop-cock making it possible to alter the pressure in one cuff while the other remains inflated. It should be noted that because the cuffs are so narrow the pressure shown on the manometer is not that exerted on the leg, but is merely a guide to how fully the cuff is inflated. When viewing the radiographs for perforating vessels their position is measured above the malleoli to assist the surgeon in their localisation at operation. Simple measurement with a ruler, on the radiograph is in-accurate owing to the magnification effect of the radiographic technique and the diverging beam of the X-rays. To overcome this difficulty a long thin strip of wood, with nail markers inserted at measured intervals, is placed alongside the limb to be examined. These nail markers will then appear on the radiograph, and be subject to the same radiographic distor-tion as the limb. Simply counting the nails will give the exact distance above the malleoli where the offending vessel can be found.

The needle used is a No. 12 hypodermic with a short bevel; or a scalp vein set may be employed. When cut-down is necessary a Hamilton-Bailey gilt cannula or a gamma sterilised polythene catheter can be used.

Departmental Preliminaries

1. A preliminary film is taken on the screen to give the correct exposure. A film of the knee region is taken since the calf and thigh exposures can then be inferred from it.

2. The procedure is explained to the patient; in particular, he is taught the Valsalva manoeuvre since this is performed just before and during each exposure.

3. In nervous patients it may be judged advisable to give further sedation, e.g. pethidine 50 mg. and promethazine 25 mg. intravenously.

4. The ankle is compressed to distend the foot veins and a suitable vein on the dorsum of the foot is selected for injection of contrast medium. In cases where there is difficulty in finding an adequate vein, bathing the foot in warm water will often help to make the veins stand out better and render percutaneous puncture easier.

5. The pneumatic cuffs (or the tourniquets) are firmly applied, one just above the ankle, the second below the knee. When there is an ulcer on the lower leg it is usually possible to apply the cuff below it; if not, the ulcer should be protected with gauze swabs and the cuff applied over it. Should this cause too much pain the cuff is put above the ulcer.

If rubber tourniquets are used the rubber should be reasonably supple so that the degree of stretching can be easily controlled. The tourniquet is fastened with artery forceps on the medial side of the leg.

Technique

The operator and assistant scrub up. The foot and ankle are liberally cleaned with hibitane spirit and sterile towels are applied around the foot and ankle, leaving only the dorsum exposed.

The short bevel hypodermic needle is attached to a polyvinyl connection and thence to a syringe of saline, the whole system being filled with saline. The selected vein is punctured directly; it is not normally necessary to use local anaesthetic. When the vein is thought to have been entered, the assistant draws back on the syringe. Provided that the puncture is satisfactory the needle is strapped *in situ* with short lengths of adhesive strapping; the connecting tubing is also strapped in the region of the toes. The assistant keeps the needle patent by gentle, intermittent injection of small amounts of saline.

In most cases it is possible to perform phlebography by this percutaneous method. Occasionally it is necessary to cut down on a vein and insert a small cannula (local anaesthesia being used). The vein over the head of the first metatarsal is usually found to be suitable.

When all is ready, the cuffs (previously inflated just enough to engorge the foot veins) are fully inflated giving a manometer reading of about 300 mm. Hg. Fifty ml. of contrast are drawn into the injecting syringe which is attached to the connecting tubing; injection is then commenced under screen control.[1]

The contrast is watched as it passes as far as the lower cuff. It should then run through into the deep system of veins. If it enters the superficial system, the injection is stopped and the lower cuff is further inflated. Should the contrast fail to pass the cuff, the pressure is slowly reduced until filling of the deep veins begins. The remainder of the contrast is injected slowly; when good filling of the calf veins to the knee is seen the patient is asked to perform the Valsalva manoeuvre and a film is taken ($14'' \times 14''$ films are used throughout). This film should include the area covered by the lower cuff so that one can later check whether any superficial venous filling is due to leakage past the cuff from inadequate pressure.

Next the leg is rotated outward and a lateral film of the calf is taken, again with the patient performing the Valsalva manoeuvre. The leg is then straightened and filling of the knee region observed. If necessary, the pressure in the 'below-knee' cuff is adjusted in the same way as at the ankle, so as to control the filling of the deep thigh vessels. Screen control enables the correct pressure to be applied.

The third film taken is an A.P. of the knee including the upper pressure cuff. A lateral film is not taken as a routine but is employed whenever an incompetent short saphenous vein is suspected. Further films are taken in the thigh. During the procedure only sufficient contrast is permitted to pass the cuffs to allow adequate films of each area to be obtained. Usually when good films of the thigh have been taken there is still a moderate quantity of contrast both in the thigh and below the knee. At this stage the fluorescent screen is centred over the groin and pelvis and the cuffs are completely deflated. Sufficient contrast then enters the groin and pelvic regions for good visualisation of the external and common iliac vessels to

---

[1] If image intensification is not available it is still possible to commence screening almost immediately since the foot can be clearly seen on the screen without preliminary adaptation.

be obtained. A radiograph of this region is taken. If necessary a second film of this area can be taken following massage of the leg and thigh. The Valsalva manoeuvre is performed by the patient before each exposure. Immediately after the last exposure has been made both cuffs are loosened, and the films are awaited. During this period 100–150 ml. of heparinised saline is infused. If they are found satisfactory the needle is withdrawn and a small pressure dressing is applied. The leg is elevated for a few minutes to empty it of contrast. After cut-down the wound is sutured and dressed.

When inspecting the films, one should look for:—

(a) the number, site and competence of the deep venous valves
(b) patency of the deep venous system; localised areas of non-filling or areas of luminal irregularity indicative of past or present thrombosis.

Inadequate or irregular filling can occasionally occur for technical reasons and when this is suspected a second injection may be made using a larger volume and paying particular attention to this area.

(c) varicosity of the deep venous system; in some cases there may be aneurysmal dilations.
(d) incompetent communicating vessels which will show superficial vein filling. Their site should be noted by counting the markers on the 'marker board'. It is essential not to overlook incompetent perforators in the adductor canal.
(e) any possible drainage veins in the base of an area of ulceration.
(f) reflux of contrast from the deep femoral vein to the saphenous vein in the groin, and from the popliteal vein to the short saphenous vein. A lateral film of the knee region may be required to show this adequately.
(g) accessory saphenous veins.
(h) the presence of any collateral channels especially proximal to an iliac vein occlusion.
(i) the patency and contour of the iliac veins. When the iliac veins have not been adequately visualised, then the examination should be repeated to show them satisfactorily. Without this the examination is incomplete.

After care

The patients may be ambulant immediately after the examination or after full recovery from the pre-medication when this is used. Sutures are removed five to six days after a cut-down.

Difficulties

1. The commonest difficulty is an inability to obstruct the superficial venous system at the ankle. This occurs in two main types of patient:

A. *Those with large painful ulcers at the ankle*. In such cases an adequate phlebogram can be obtained even though compression of the superficial veins is not employed. Both the superficial and the deep venous systems fill; any communicating veins feeding the ulcerated area will be outlined and their superficial tributaries will be visualised. It is possible to measure their distance above the malleoli and so assist the surgeon in his search for them at operation.

B. *Those with gross oedema.* In some oedematous limbs, even when considerable compression is applied, the superficial veins always fill before the deep system. This may be secondary to old deep vein thrombosis with consequent alteration in the haemodynamics. In these cases both deep and superficial filling will occur and it becomes difficult to determine the competence of any communicating veins which are demonstrated. In the great majority of such cases, however, the clinical problem is related not to the communicating vessels but to abnormalities of the deep venous system such as incompetent valves and varicosities. These latter will be demonstrable in spite of inadequate compression.

2. Another technical difficulty is due to the production of multiple haematomata from failed attempts at percutaneous puncture. When, eventually, successful puncture is achieved and compression applied, extravasation of contrast can take place through any of the sites of earlier venous puncture. For this reason it is wise not to persist in attempts to obtain a percutaneous puncture if there is difficulty but to proceed to a cut-down at an early stage. With experience one becomes able to recognise those cases likely to give rise to difficulty.

3. Blockage of the needle can occur if the assistant does not maintain continual irrigation with saline. We have found that when it becomes impossible to inject saline through the needle, it is often more satisfactory to introduce a new needle than to attempt to clear the obstructed needle.

Complications.—Complications are uncommon, but phlebitis may occur; it is most often due to infection but can also be due to the trauma of puncture or to the contrast. It is particularly likely to occur when there is already clot formation in the veins and particular care should then be taken to drain the veins of contrast, using limb elevation and irrigation with 250 ml. heparinised saline. Treatment with antibiotics and anti-coagulants may be necessary.

References

GRYSPEERDT, G. L. (1953). *Br. J. Radiol.*, **26**, 329–338.
Dow, J. D. (1951a). *Br. J. Radiol.*, **24**, 182–192.
Dow, J. D. (1951b). *J. Fac. Radiols.*, **2**, 180.

HALLIDAY, P. (1968). *Br. J. Surg.*, **55**, 3, 220.
CRAIG, O., and SNELL, M. (1966). *Br. J. Surg.*, **8**, 53, 668–672.

Further Reading

DODD, H., and MISTRY, M. (1958). *Postgrad. med. J.*, **34**, 437–442.
DEWEESE, J. A. (1958). *Surgery*, **44**, 390–397.
MOORE, T. C. (1956). *Archs Surg., Chicago*, **72**, 122–135.
ARNOLDI, C. C. (1957). *Dan. med. Bull.*, **5**, 65–71.

DOHN, K. (1958). *Acta Radiol. Stockh.*, **50**, 293–309.
GULLMO, A. (1956). *Acta Radiol.*, **46**, 603.
GULLMO, A. (1957). (*Ibid.*), **47**, 119.
GULLMO, A. (1963). *Br. J. Radiol.*, **36**, 431, 812.

APPENDIX

Trolley setting

Basic vascular trolley setting (p. 191) with the following additions:—

Sterile—upper shelf

Short bevel hypodermic No. 12 needle.
Hamilton Bailey gilt cannula (small).
Cut-down set:
   1 pair scissors.
   2 pairs mosquito forceps.
   1 small needle holder.
   1 small non-toothed dissecting forceps; 1 toothed dissecting forceps.
   1 aneurysm needle.
   1 blade-holder with No. 15 blade (if not already included in setting).
   Catgut size 2/0 and linen thread. 1 Bonney half-circle cutting needle.
A 50 ml. syringe may be of value.

Lower shelf

Contrast medium—6–8 ampoules of Urografin 60% or vials of Conray 280 in a bowl of hot water. Ampoules with catgut sutures size 2/0.

Also required

Pair of narrow inflatable cuffs (St. Mary's Hospital pattern), clock-faced manometer with three taps and a hand bellows (obtainable from V. J. Millard). *N.B.*—If the I.V.U. compression apparatus described on p. 46 is available, this will provide a suitable manometer and hand bellows.

Alternatively, two rubber tourniquets are employed with two pairs of artery forceps. As already noted the rubber should be supple.

Various sizes of Abbott Butterfly scalp vein sets should be available. Generally the 21G thin-wall size is suitable.

CHAPTER 31

SUPERIOR VENA CAVOGRAPHY

In this examination the aim is to outline the veins of the thoracic inlet and the superior vena cava. The equipment used has varied from a large needle on a syringe to bilateral catheters in the arm veins. As with other types of venography, difficulties in interpretation may arise if venous filling with contrast is incomplete, since the influx of unopacified blood can cause confusing defects in the contrast. For this reason we recommend that every effort be made to achieve injection of contrast into both arms at once. Failing this the injection of each arm in turn may be carried out (Howard and Pick, 1961) and the resulting films used to obtain a composite picture of the state of the veins. If this should be necessary it is of value to occlude the venous return from the arm which is not being injected, using a sphygmomanometer cuff.

Contra-indications.—Active heart disease, e.g. recent myocardial infarction.

Ward preparation

1. The patient has no food or fluid during the 4 hours prior to examination.

2. If the arms are oedematous the ward is asked to encourage the patient to keep them elevated as much as possible.

3. Scopolamine 0·4 mg. is given intramuscularly half an hour before the examination.

4. The patient empties his bladder before coming to the X-ray department.

5. The patient wears a loose-fitting gown with short sleeves.

Department preparations

Medium length, i.e. 50–60 cm. polythene catheters are prepared. OPP 205 catheter tubing will normally allow reasonably rapid injection of contrast but for really forceful injection, particularly when only one arm can be catheterised, the best combination is OPP 240 drawn out to fit the plastic PE 160 leader (p. 237) with 2–3 side-holes eccentrically placed towards the tip.

Department preliminaries

1. The patient is examined. The points to be determined are:—

(a) General signs of S.V.C. obstruction, to assess its severity.

(b) The state of the arm veins: these are usually distended and so may be easy to feel but if there is oedema as well as venous distension they may be impossible to palpate satisfactorily. A modified cut-down technique is then used.

(c) Whether the patient can tolerate lying flat for a period. If he cannot the catheterisation of the veins is best performed on a tilting table with the feet lowered (footstand on the table) about 20–30°. This will normally provide relief from the discomfort of lying flat but if even this cannot be tolerated it should be possible to catheterise the veins with the patient sitting and his arm over a table. We have never found this necessary.

2. The examination is explained to the patient warning him that the injection will cause a suffusion of warmth in the neck, upper chest and head.

3. The preliminary film is taken on the film changer. This should include the lower part of the neck, the side-walls of the chest and the whole of the heart shadow. (Although a film changer is more convenient, multiple films taken on the Bucky are usually adequate.) The centring point for this exposure is marked on the skin together with the edges of the area covered. If the patient is to be tilted during the catheterisation he is now lifted onto a tilting table and his feet are lowered.

Summary of the procedure

1. Plain film of chest and neck.
2. Catheterise both basilic veins by the Seldinger method, using a modified cut-down technique if necessary.
3. Join both catheters to a syringe of contrast via a Y-piece (or have two operators, one injecting each arm).
4. Inject contrast and take multiple films, 4 at least.

Procedure

Contrast.—45% Hypaque or 280 Conray is used, 25 ml. into each arm.

Equipment.—1. *Needles.*—As already mentioned it is possible to carry out this examination with a relatively wide needle, e.g. 18 gauge. To reduce movement during forced injection it is advisable to link the syringe and needle by an arteriographic connection (p. 194). Even then the risk of extravasation is not inconsiderable and if a needle of this size can be inserted into the vein it should be possible to use the 18 G arterial cannula so that the fine PE 160 leader can thus be introduced.

2. *Cannulae.*—Another approach is to use the 18 gauge arterial cannula with a short leader as described for femoral arteriography. This is an improvement on a needle but is still less satisfactory than catheterisation.

3. *Catheterisation.*—The best method at the moment is Seldinger catheterisation using OPP 205 tubing or OPP 240 drawn out over the PE 160 leader.

Technique

1. With the patient settled comfortably a sphygmomanometer cuff is applied to the arm and if necessary inflated to a suitable pressure further to engorge the veins. The elbow is extended over a pad and the skin prepared. Towels are draped over the forearm and abdomen.

2. Whenever possible the basilic vein is used as filling of the subclavian vein is much less reliable from the cephalic. A skin wheal of lignocaine is raised on an appropriate point below the point of intended venepuncture and the skin is nicked with a small scalpel.

3. Puncture of the vein is then carried out using the 18 gauge arterial cannula or a Seldinger PE 160 needle-cannula. When the Seldinger needle-cannula is employed the central needle is drawn out leaving the cannula in the vein; the leader is then passed through the cannula. Thereafter the technique is similar to that for catheterisation of the arteries. It is not necessary to control the catheter position precisely but the tip should be placed about 10–15 cm. from the elbow, i.e. leaving about 35–45 cm. outside. A pressure pad is taped over the puncture site and the catheter is fastened just below this with adhesive plaster. The catheter is flushed intermittently with dextrose while puncture of the vein in the opposite arm is carried out in the same way.

4. Should there be any difficulty in defining the vein for percutaneous puncture, from oedema or other causes, a modified cut-down is performed. In this method the skin over the vein is infiltrated with lignocaine wherever the vein can best be felt. A transverse incision is then made across the line of the vein as before a conventional cut-down. Blunt dissection—opening out the tips of a pair of scissors—is used to expose the vein. This is then punctured directly and the Seldinger leader and catheter passed into it. Should it be difficult to steady the vein a controlling loop of catgut passed around it may be found useful to draw it downwards and stretch it. Occasionally a full cut-down may be necessary.

5. When both catheters are in position they are linked—using an arteriogram connection if necessary—to a Y-connection and thence to a dextrose drip (patients who have been semi-recumbent during catheterisation are now transferred to the cassette changer). When all is ready a syringe containing 50 ml. of 45% Hypaque or 280 Conray is attached (simultaneous injection of each arm by two operators is, of course, equally satisfactory).

6. The patient is now told to breathe in slowly and to hold his breath. Injection of contrast is made during and after inspiration and the first film is exposed towards the end of the injection. At least three further exposures are made as rapidly as possible.

7. While awaiting the films the patient may be propped up. Where the cut-down technique has been employed sutures may be put in the skin and around the vein and tied loosely.

8. When the films have been seen and are thought to be satisfactory the catheters are removed, any sutures are tied firmly and pressure bandages are applied.

Reference

HOWARD, N., and PICK, E. J. (1961). *Clin. Radiol.*, **12**, 290.

APPENDIX
Ward instructions
1. If the arms are oedematous keep them elevated as much as possible.
2. Nothing by mouth for 4 hours before the examination.
3. Ask the patient to empty the bladder before coming to the X-ray department.
4. Give scopolamine 0·4 mg. i-m. at . . . . .a.m./p.m.

Trolley setting

Basic vascular trolley setting with:—
1. Additions for Seldinger catheterisation.
2. Y-connection (M.D.) and pressure injector *or* second 30 ml. syringe for injection of contrast.
3. Cut-down set—see Appendix to Chapter 29, p. 256.

Also needed

Sphygmomanometer—to occlude venous return from the opposite arm if only one side is injected.

CHAPTER 32

INFERIOR VENA CAVOGRAPHY

THE aim of this investigation is to outline the entire extent of the I.V.C. while avoiding pitfalls in interpretation due to inadequacy of method. There are several techniques for carrying out this procedure but that evolved by Helander and Lindbom (1955) is particularly satisfactory. It involves percutaneous puncture and catheterisation of both common iliac veins by the Seldinger method using two polythene catheters.

Contra-indications
1. Evidence of femoral vein thrombosis.
2. Infection of the groin.
3. Absent pulsation of a femoral artery (these arteries being vital as landmarks).
4. If the patient is receiving anticoagulant therapy especial care is needed to prevent haematoma formation.

Special points.—General anaesthesia is unnecessary other than in exceptional circumstances. A tilting table is not required. The patient should be admitted to a ward on the previous night and kept in for the night following the examination.

Ward preparation.—Both groins must be shaved. The colon is 'cleared'. The patient is starved for 6 hours.

Departmental preparation.—All the equipment necessary for catheter aortography will be required (p. 235). Two polythene catheters are prepared, about 50–60 cm. in length. Size OPP 205 tubing and the corresponding needles were employed by Helander and Lindbom (1955) but we have found OPP 240 catheters to have advantages, in allowing rapid injection of a large bolus of contrast; side holes must be made near the tip. The two catheters are joined by a Y-connection which in turn is attached to a pressure injector—e.g. Talley pump. (The use of a Y-connection is merely for convenience, i.e. one syringe will be able to inject both veins, but just as good a result will, of course, be obtained with separate syringes and two operators, each injecting 30 ml. of contrast.

Departmental preliminaries.—Examine the patient and explain the procedure to him in simple terms, mentioning that during and after the injection he will feel a sensation of warmth in the abdomen and that he must keep absolutely still during this time. He should be taught how to hold his breath without straining, by closing his lips midway between inspiration and expiration while a clamp is placed on his nose. The Valsalva manoeuvre is not ordinarily used—as it sometimes causes uneven filling and subsequent mistakes in interpretation. It is employed if the origins of tributaries, such as the renal veins, are to be shown.

Premedication.—The premedication is supplemented if necessary.

Positioning the patient.—Films may be taken in one of two positions.

1. A.P.—patient supine.
2. Oblique—patient supine, left side raised 45°.

Both projections are satisfactory but the simple supine position gives films which, anatomically, are somewhat easier to interpret. The position chosen depends, in fact, on the particular problem under investigation. The oblique view is sometimes necessary to show impressions by enlarged retroperitoneal lymph nodes. Normally the A.P. projection should be used and, if this is not satisfactory, the patient is then placed in the oblique position for the second series.

A preliminary 17″ × 14″ film of the abdomen and pelvis is taken; this will usually cover the area under investigation quite satisfactorily without having to move either the patient or the film carrier during the injection.

Procedure

Contrast medium.—Unlike aortography a large volume of a relatively low concentration of medium is injected (i.e. 60 ml. of Urografin 60% Hypaque 45% or Conray 280). This results in a really good mixture of blood and contrast medium, hence avoiding subsequent errors in interpretation; only in the obese or very large patient may the higher concentrations of medium be required, (i.e. 76% Urografin or Conray 420).

Technique

The preliminary film having been checked for position and exposure, the right femoral artery is palpated with a finger of the left hand. The puncture is made immediately adjacent and medial to the palpating finger with the point of the needle towards the head of the patient. When local anaesthetic has been injected and the skin incised, the arterial cannula is plunged in fairly deeply so that it passes right through the underlying femoral vein—next the cannula is withdrawn very slowly (as with a femoral artery puncture) until there is a fairly fast, even, non-pulsating flow of venous blood. Lower the butt end of the cannula a little and feed in the leader. When 8–10 cm. of leader have been introduced, withdraw the cannula and introduce the catheter for 10–12 cm. only, and fix it to the skin of the patient with strips of sterile autoclave tape. Now catheterise the left femoral vein in a similar manner. (If there is difficulty in introducing the leader *never* try to force it, withdraw the cannula 2 mm. and try again; if it still fails to pass, push the needle-cannula through the vein again, withdraw slowly and repeat the performance. Should no venous flow be obtained this may be because either the venous pressure is too low or the vein is thrombosed. It is then advisable to attach a flexible connection to the cannula before it is withdrawn. An assistant aspirates gently as the cannula is pulled back. Alternatively, the patient may be asked to perform a Valsalva manoeuvre as the cannula is being withdrawn. Continued failure to puncture the vein will necessitate trochanteric puncture (p. 275) to show the veins on that side. If one side is successfully catheterised and if the examination is mainly intended to show the cava itself the catheter is placed at the junction of the common iliac veins and a pressure injection of 50 ml. Conray 420 is made. The catheter should then be an O.P.P. 240 with three side holes and the Valsalva manoeuvre is of value in filling the mouths of the iliac veins.

When only one side can be punctured and when it is thought important to examine the opposite side this can be done by carefully advancing a curved green Kifa catheter through the 'good' side up to the bifurcation and down the opposite iliac vein with intermittent injection of contrast to ensure that it is not about to encounter any clot.

When both catheters are well placed in the veins, withdraw the leaders, allow a little blood to flow back along each catheter, clear them with dextrose and attach them both quickly to the Y-piece; one syringe of dextrose will now keep both catheters 'clear'.

Fill the metal injection syringe with 60 ml. Urografin 60% (or 45% Hypaque or Conray 280). Check with the radiographer that all is ready. Attach the Y-connection to the injection syringe. Warn the patient to stop breathing, and immediately inject all the medium within 4 seconds (15 ml./sec.). Make the first exposure when half the medium has been injected and take (if possible) up to 10 films at one-second intervals. The number of films taken will depend upon the particular problem under investigation and the type of changer available. In fact, a useful examination can ordinarily be obtained with only 2 or 3 films changed by hand in the Bucky tray.

After the first series, leave the catheters *in situ* until the films have been checked and then repeat the examination, several times if necessary, until entirely satisfactory films have been obtained. Then withdraw the catheters and compress the puncture sites. When compression has been exerted for about 5 minutes the puncture sites are inspected for a minute and if no bleeding is occurring small plaster dressings are applied. Pressure dressings are seldom needed after venography except when the patient is on anti-coagulants.

Pelvic Venography.—If it is proposed to investigate the pelvic veins rather than the inferior vena cava itself, the technique just described should be modified as follows:—

1. Just before the injection apply abdominal compression to the right of the umbilicus using a Bucky band and a block or a urographic compression apparatus. Make quite sure that the femoral artery pulse is not occluded (release pressure as soon as the exposures are completed).

2. The *full* Valsalva manoeuvre should be used during the exposures. (This must be taught to the patient before the procedure is begun.)
3. The X-ray tube is centred on the pelvis and tilted so that the beam is directed 10° towards the head. Apart from these minor modifications the technique is identical.

Renal venography.—The renal veins may be investigated for a number of reasons, notably the presence or absence of tumour or clot. Indirect evidence of the patency of the veins as provided by negative shadowing in the caval contrast column due to the inflow of blood from the renal veins is not wholly reliable. To prove that the renal veins are normal requires either direct catheterisation or their filling via a renal artery injection of contrast. There is usually little technical difficulty in catheterising the veins though the shape of catheter required for the two sides may be somewhat different—see Fig. 42, p. 241. The main difficulty with renal venography arises in filling the venous system completely since the continuous inflow of blood from the kidney tends to wash out any contrast almost as soon as it is injected and even pressure injection will seldom produce really complete filling. To overcome this one can inject adrenalin into the renal *artery* on the corresponding side after selective catheterisation (Olin and Reuter, 1965). 0·2 μg. per kilogramme of adrenalin are injected diluted in saline. This causes temporary constriction of the renal vessels so allowing the venous injection to be made without being immediately washed out. In fact, if this approach is contemplated it is better to catheterise the renal arteries first and inject contrast directly so as to produce opacification of the veins on the late films. This will then prove that the smaller renal veins are not obstructed and may obviate the necessity for venous catheterisation. If catheterisation is undertaken it can be made in the assurance that the renal veins are patent and with the assistance of adrenalin should this be required.

Composite examinations of the inferior vena cava and its tributaries.—It is now becoming common for investigation of the veins of the legs and trunk to be undertaken in the search for a source of pulmonary emboli. This entails venography of the deep veins of the legs together with pelvic venography and inferior vena cavography. It is usually best to start with the deep vein examination of the lower limbs using either puncture of a superficial vein on the foot with tourniquets on the ankle and knee as described in Chapter 29 or puncture of the medial malleolus as described in Chapter 33. Films are taken as the contrast travels up the leg to the groin and it can then be seen whether the femoral veins are patent or not. If patent they are catheterised in the usual way. If one or both are not patent it is then necessary to undertake intraosseous venography, injecting the contrast into the greater trochanter. Clearly it is only possible to do the latter examination if the patient is under a general anaesthetic. If this is not possible it may then be necessary to pass a catheter from the opposite side and down into the iliac vein on the side which cannot be punctured.

With this method there is obviously a risk of displacement of clot and it should be undertaken only with caution and only by those who are experienced in venography.

Complications and after care.—Apart from local haematomata, complications are very rare provided the operator is gentle and never attempts to force the leader into the vein.

Sometimes the femoral artery is punctured in error. This is of no consequence, the error will be readily apparent from the type of flow and colour of blood. It is treated by compression.

Thrombosis of the punctured veins is very unusual but it is *imperative for all these patients to be up and walking about directly after the examination* in order to prevent this rare, but serious complication from taking place.

References and further reading

Venography of the I.V.C.
HELANDER, C. G., and LINDBOM, A. (1959). *Acta Radiol.*, **52**, 257.

Retrograde Pelvic Venography
HELANDER, C. G., and LINDBOM, A. (1959). *Acta Radiol.*, **51**, 401.

Pelvic phlebography and cavography
LINDBOM, A. (1960) in *Modern Trends in Diagnostic Radiology* (Third Series) (Butterworths, London) iii.

Renal Venography
OLIN, T. B., and REUTER, S. R. (1965). *Radiology*, **85**, 1036.

APPENDIX

Ward instructions

As for transfemoral catheter aortogram.

After care

1. The patient should be up and walking as soon as possible after returning to the ward. If a general anaesthetic has been given, raise the legs every 15 minutes until the patient is able to get out of bed.
2. Quarter-hourly inspection around the dressings to make sure no haematoma formation has occurred for 2 hours; then half-hourly inspection for 2 hours.
3. Half-hourly pulse for 4 hours; then 4-hourly T.P.R.

Trolley setting

As for transfemoral catheter aortogram with the addition of a Y-connection (M.D.). A suitable pneumatic thigh cuff is advisable to occlude venous return from the opposite leg should it prove impossible to catheterise both femoral veins (obtainable from Willen Bros., 44 New Cavendish Street, London, W.1).

CHAPTER 33

SPLENOPORTOGRAPHY
(Percutaneous spleno-portal venography)

THE aim of this examination is to delineate the splenic and portal venous system together with the liver 'pattern' and, if possible, the hepatic veins as well, by means of a percutaneous puncture of the splenic pulp. Carried out correctly this should be a relatively short and simple procedure carrying minimum risk or discomfort to the patient. The radiologist should visit and examine the patient in the ward and chat briefly about the examination in a reassuring manner. He should familiarise himself with the history, examine the abdomen and percuss and palpate the spleen.

Contra-indications.—When there is:—

1. *Tendency to bleed.*—In the presence of liver disease this is most likely to be due to deficient prothrombin formation and the prothrombin time estimation and platelet count are always carried out.

2. *Deep jaundice.*—Even when the prothrombin time is normal haemorrhage appears to be more likely to occur in the presence of deep jaundice. (Turner *et al.*, 1957).

3. *Any condition predisposing to splenic rupture.*—This includes such diseases as malaria and glandular fever.

Special points.—The prothrombin time must be estimated before the examination. If the prothrombin time is prolonged vitamin K is given. Only when the prothrombin level is within safe limits may the investigation be carried out. (Figures of such levels cannot be given since methods of estimation vary with each laboratory; the matter must be discussed with the physician in charge of the case.) The platelet count should exceed 100,000/cu. mm. These points must be carefully checked since otherwise dangerous bleeding may occur after the examination. The examination must never be carried out on an out-patient, the subject being admitted to hospital for the nights before and following the examination.

A general anaesthetic is not usually necessary in an adult but in children and nervous adolescents general anaesthesia is essential and in that case one should make sure that the anaesthetic department is informed well in advance.

Ward preparation of patient.—This is identical with that required for a lumbar aortogram. (See Chapter 28, page 211). If ascites is present the ascitic fluid should be tapped on the morning of the examination.

Premedication.—If desired papaveretum and scopolamine may be used in most cases. Opiates are, however, broken down by the liver and in severe liver disease pethidine is safer. The short acting barbiturates are also broken down in the liver and are similarly contra-indicated (Sherlock, 1958).

Equipment required.—A catheter-over-needle (Seldinger, 1957) is preferable for this examination. The commercially available Teflon needle-catheters are more than adequate for this purpose. The catheter itself need not be unduly long, 7 cm. usually being adequate but it is convenient to have a length of catheter outside the patient and a 10 cm. (4 inch) catheter is commonly used. The examination can be carried out using a needle alone but this increases the risk of injury to the spleen; furthermore, the needle must be withdrawn immediately after injection so that if a second injection is required a second puncture is necessary.

Contrast medium.—60% Urografin, 45% Hypaque or 280 Conray is used.

X-ray equipment required.—An automatic changer capable of taking 16 films is an advantage. When a cassette changer with a limited capacity is used extra cassettes should be available so that continued exposures can be made after the first rapid series (see below).

If the catheter method of injection is employed it should be possible to screen the patient with an image intensifier before the films are taken; a small test injection of contrast can then be watched, thus enabling the catheter position to be accurately determined.

Departmental preliminaries.—Examine the patient again and percuss and palpate the spleen. Mark the 10th left rib indelibly on the skin and tape a lead marker over the proposed site of puncture.

Explain the procedure to the patient in simple terms—warn him that he will feel a prick in his left side followed by a warm sensation in the abdomen and accompanied by a great deal of noise when the exposures are made and the films changed. Instruct him in shallow breathing and make him practise sudden apnoea on command: when all these instructions are clearly understood, position the patient supine with the hands behind the head or over the upper chest.

The position for examination.—The supine position has usually been adopted for splenoportography but has the disadvantage that the relatively heavy contrast medium may run into the most dependent part of the portal system and may fail to fill varices. In view of this it may well be that the routine use of the prone position is preferable as suggested by Moskowitz et al. (1968). Our own experience is insufficient to allow a firm recommendation but it is clear that even if the prone position is not used routinely it should be employed if examination in the supine position fails to show varices when their presence seems likely. It is of interest that Moskowitz and his colleagues found no difference in the splenic pressure between the supine and prone position.

Preliminary films.—A single A.P. film (17" × 14") is taken to test exposure and field size; the field should include both spleen and liver and should be wide enough to demonstrate possible oesophageal varices and lower abdominal collateral veins (centre to the xiphoid). It will also be useful for delineating the spleen relative to the marker (previously placed over the puncture site). Sometimes the splenic outline is not apparent but this need not affect the success of the puncture. The patient should hold his breath 'mid-way' during the act of shallow breathing for this and subsequent films when instructed to 'stop breathing'. It is most important that that patient should hold his breath at the same stage of respiration on each occasion otherwise the relationship of the spleen and marker may alter considerably, so confusing the position for subsequent puncture. This should be emphasised to the anaesthetist when the examination is carried out under general anaesthetic.

Procedure

Puncture site.—The position of the spleen and its relation to the skin marker can usually be assessed from the preliminary film, and use of this information in combination with the methods described below increases the likelihood of correct insertion of the needle.

The smaller the spleen the further back should the puncture be made and when the spleen is impalpable puncture is below the 10th rib in the posterior axillary line. To make this easier the left side is raised 20–30°. The needle is directed slightly cephalad and angled about 10–15° anteriorly. Once the needle has been satisfactorily introduced the patient lies flat (supine).

When the spleen is considerably enlarged puncture is made where it is apparently closest to the abdominal wall (by palpation) in the mid or anterior axillary line. The needle is directed about 10–15° cephalad and with more anterior puncture is directed somewhat backwards.

If the spleen is palpable but not markedly enlarged puncture is made in the 8th or 9th intercostal space in the mid axillary line, directing the needle cephalad as described above.

Local Anaesthesia.—The lead marker is removed from the skin, which is prepared in the usual way. The site for puncture is finally decided after inspection of the preliminary film. The skin is then infiltrated with 1% lignocaine and then the deeper tissues for a depth of 2–3 cm., i.e. roughly down the splenic capsule. A 'nick' is made in the skin with a scalpel blade. (When the needle reaches the region of the spleen a 'grating' may be felt as the point scratches the capsule. This gives an exact indication of the depth of the spleen which is of value during subsequent puncture.)

Method of puncture.—The needle-catheter is advanced through the skin puncture and into the tissues for about 2–3 cm., i.e. down to the region of the splenic capsule. A 'grating' may again be felt, confirming that the needle tip lies close to the spleen. (At this stage it is an advantage with patients who are breathless to withdraw the needle slightly and tell the patient to hyperventilate for about 20 seconds.)

The patient is then told to stop breathing in mid-inspiration. The needle-catheter is directed as described earlier and plunged in to a depth of 5–6 cm. The needle is quickly withdrawn leaving the catheter *in situ*. If the pulp of the spleen has been entered satisfactorily there will be a regular drip of venous blood issuing from proximal end of the needle. It is as well to support the butt of the catheter on sterile pads. The patient is now told to breathe in a shallow manner. A polyvinyl connection and a syringe of saline are attached (note that the male fitting of the connection should be Luer-Lok) and a small quantity of saline is injected. The flow-back of venous blood mentioned above is an important sign of successful puncture and *without it the full injection of contrast should not be made*; it is again checked before attaching the syringe of contrast.[1] (When screen control of the catheter position is possible, a small (5–6 ml.) injection of contrast is made and watched before transferring the patient to the cassette changer.)

Pressure measurement.—At this stage the splenic pulp pressure should be recorded. A two-way tap is attached to the arterial connection. A glass manometer is connected to the side-arm of the tap and is filled with saline. The tap is then turned to connect the catheter and manometer, the zero point of the manometer being held at the level of the couch top. When the saline level stops falling a reading is made. This is repeated after injection of saline to clear the catheter and refill the manometer. A third reading is made after further injection of saline but this time the level is allowed to rise from the bottom of the manometer. A mean of the three readings is taken and about 12 cm. deducted to allow for the position of the zero point below the point of puncture. The normal range is 100–220 mm. of saline.

The injection.—Check that the radiographer is ready. Warn the patient that he will experience a warm feeling in his abdomen—instruct him clearly to 'stop breathing' and at once inject 50–100 ml. of contrast by hand (45% Hypaque, 60% Urografin or Conray 280) as quickly as possible.

Films to be taken (A. P. plane only).—The first films should be taken as the injection starts—this is very important.

(a) If an automatic changer is used take one film a second for ten seconds, then one film every three seconds for a further six films (a total of sixteen films).

(b) If a hand changer is used take the maximum number as rapidly as possible, after this continue to insert and expose cassettes one by one up to 3 or 4 more. It may be necessary to repeat the 'run' as soon as the films are viewed, varying the timing as required.

In practice eight films at two-second intervals will usually be found to be adequate but the undoubted advantage of the automatic changer and its multiplicity of films over a long period is that it is then possible to obtain films of the portal vessels, the 'hepatogram' and the hepatic veins (Bergstrand, 1961).

---

[1] With a simple needle puncture a much more rapid technique is required. As soon as the drip of venous blood is seen, the polyvinyl connection and syringe of contrast are attached and injection made. If possible the patient should not breathe from the introduction of the needle until its withdrawal following injection. Minor delays may make it necessary to allow shallow breathing which will produce an up and down rocking of the needle in a direction opposite to that of the diaphragm.

A catheter may be left *in situ* and gently perfused with saline until the films have been developed and viewed. A repeat series is then obtained if necessary but if all is well quickly withdraw the catheter and cover the puncture site with collodion. When a needle alone has been used it is withdrawn as soon as the injection is completed; repuncture is undesirable if it can be avoided but may be necessary if the first run is totally unsuccessful.

The patient should feel only mild discomfort from the injection. If severe discomfort is experienced the injection is probably extra-splenic. This will be confirmed by the complete cessation of flow of blood after injection and will be immediately apparent on the monitor. The catheter is then withdrawn. Repuncture is not attempted until the films have been viewed:

1. because enough of the injection may have passed into the splenic vein to give diagnostic information;
2. because the optimal site for repuncture may be determined by observing where the initial puncture lay.

The spleen should not be punctured more than twice at any one examination.

After the injection.—The patient should be lifted carefully off the table and in the ward should lie as still as possible for the ensuing 4–5 hours and a careful record kept of pulse and blood pressure. He should not be allowed out of bed for at least 12 hours.

Complications and hazards

1. *Haemorrhage* from the puncture site in the spleen is seldom serious enough to produce signs of intraperitoneal bleeding—being reported in less than 2% of cases. The radiologist can make a major contribution to avoiding sequelae by strict observance of the details and precautions noted above. (If a cannula or a needle is used without resort to the catheter technique, this is liable to be a more traumatic procedure, although in experienced hands the results are very satisfactory. An excellent account is given in Bergstrand's paper (1961) among others.)

2. *Extra-splenic and subcapsular injection of contrast medium.*—This most commonly occurs when there is no splenic enlargement—it causes some discomfort which lasts from 5–30 minutes but may, with large injections, last several hours. With present-day contrast media there are unlikely to be serious sequelae.

Alternative methods

A number of other approaches to opacification of the portal venous system are possible. Those involving operative catheterisation of a jejunal vein or injection of contrast into a haemorrhoidal vein are outside the scope of this book. An alternative method which may be required either because of difficulty in puncturing the spleen or because the spleen has been removed, is arteriovenography, (Kreel and Williams, 1964). In this method either the coeliac axis or superior mesenteric artery are catheterised and a fairly large volume of contrast, e.g. 30 ml. of Conray 420, is injected either by hand or using a pressure injector. It is essential in this method to obtain films in the late stage of filling; those taken between about 8 and 20 seconds after injection of contrast are usually the most informative. Although the detail is inferior to that obtained at splenic puncture it is usually sufficient to show whether or not the portal vein is patent.

References

BERGSTRAND, I. (1961). In *Angiography*, vol. 2, Ed. Abrams, H. L. (J. and A. Churchill, Ltd., London).

KREEL, L., and WILLIAMS, R. (1964). *Br. med. J.*, **2**, 1500.

MOSKOWITZ, H., CHAIT, A., MARGULIES, M., and MELLINS, H. Z. (1968). *Radiology*, **90**, 1132.

SELDINGER, S. I. (1957). *Acta Radiol.*, **48**, 93.

TURNER, M. D., SHERLOCK, S., and STEINER, R. E. (1957). *Am. J. Med.*, **23**, 486.

SHERLOCK, S. (1958). *Diseases of the Liver and Biliary System.* (Blackwell Scientific Publications, Oxford.)

Further reading

STATTIN, S. (1959). *Acta Radiol.*, **52**, 353.

BERGSTRAND, I. (1961). As above.

STEINER, R. E., SHERLOCK, S., and TURNER, M. D. (1957). *J. Fac. Radiols.*, **8**, 158.

APPENDIX

Ward instructions

As for direct puncture aortogram but omit pubic shave. The blood group *must* be established.

After care

The patient is kept still in bed for 4–5 hours with quarter-hourly pulse and half-hourly blood pressure recordings. He is kept in bed for at least 12 hours with a 4-hourly T.P.R.

Trolley setting

Basic vascular setting with a Teflon needle-catheter (4 inch, 18 gauge; serial no. 01–0048, Becton Dickinson (G.U.)). A manometer as used for lumbar puncture is also needed, together with a 2-way tap.

CHAPTER 34

INTRA-OSSEOUS VENOGRAPHY

(Ian Isherwood, M.B., Ch.B., F.F.R., D.M.R.D.)

INTRA-OSSEOUS venography was first suggested by Erhardt and Kneip (1945). Since then injections have been made at many sites to demonstrate various aspects of the venous circulation and Schobinger (1960) has suggested more than fifteen injection sites in man. The basis of the method is that when contrast is injected into bone marrow it is rapidly taken up by the local venous drainage. This approach may therefore be used to replace or supplement conventional venography, e.g. lower limb phlebography or inferior vena cavography. It may also be employed to demonstrate venous networks which are inaccessible to other forms of examination, e.g. the lumbo-azygos system and the vertebral venous plexus.

The commoner sites of injection are:—

> Vertebral spinous processes.
> Malleoli.
> Greater trochanters.
> Ribs.

One of the most useful sites of injection, first suggested by Fischgold *et al.* (1952), is a vertebral spinous process which provides an easily accessible bony prominence with a rich draining venous plexus. The applications of intra-osseous injection at such a site are manifold and vary according to the level employed. This procedure will be considered in detail. The general principles of the method are applicable at other sites and the interested reader should consult Schobinger's (1960) book for a discussion of the technique and interpretation in various regions.

SPINAL INTRA-OSSEOUS VENOGRAPHY

A. General Application

1. Demonstration of:
   (a) Vertebral venous plexus.
   (b) Lumbo-azygos venous system.
2. Investigation of obstructive lesions in the spinal canal by demonstration of compression, displacement or circulatory disturbance of the internal vertebral venous plexus.
3. Early confirmation of the presence of metastases in the vertebral bodies by demonstration of epidural venous occlusion.
4. Investigation of spinal angiomata and associated vascular anomalies.

B. Specific Applications

1. *Injection of a cervical spinous process.*—At this level the usefulness of the method is limited to demonstration of the external vertebral plexus—occasionally required in the investigation of changes in the haemodynamics of cerebral venous drainage. The cervical internal vertebral plexus may be demonstrated by injection of a cervical vertebral body (Greitz *et al.*, 1962).

2. *Injection of a thoracic spinous process*
   (*a*) Investigation of diseases of the chest and mediastinum by demonstration of compression or displacement of the azygos venous system.
   (*b*) Investigation of portal hypertension by demonstration of changes in calibre and degree of opacification of the azygos vein.

3. *Injection of a lumbar or sacral spinous process*
   (*a*) Investigation of low back pain by demonstration of compression or displacement of the internal vertebral venous plexus or by the demonstration of epidural varicosities.
   (*b*) Inferior vena cavography.

Contra-indications

1. History of allergy or known idiosyncrasy to opaque-medium—but see Chapter 3, p. 15.
2. Local infection.
3. Haemophilia or other 'bleeding' tendency.

Ward Preparation—

1. The meal preceding examination is omitted, i.e. food and fluid are withheld from the patient for 4–6 hours prior to examination.
2. In-patient bowel preparation is advisable when inferior vena cavography is contemplated (see Chapter 1, p. 5).
3. Premedication—Morphine 10 mg. or Omnopon 20 mg. i.-m., 45 minutes prior to examination.

The patient is usually admitted for one night following the procedure.

Ward Preliminaries.—The patient should be examined on the ward by the radiologist with particular reference to the site of proposed injection. The procedure should be explained to the patient and warning given of possible discomfort during injection in cases who are to be examined under local anaesthesia.

Department Preliminaries.—The patient is placed in the prone position on the X-ray table for injection of thoracic, lumbar and sacral spinous processes. A sitting position is advisable for injection of a cervical spinous process.

Two preliminary films are taken (15″ × 12″ is the size normally employed).

1. A lateral film with horizontal ray and a fixed grid to assess the inclination of the vertebral spinous processes.
2. A postero-anterior film, preferably with a Potter-Bucky diaphragm.

It is important to observe the inclination of the spinous process on the preliminary films since the bone marrow needle and stilet must be introduced into it without transfixing it. The cervical and thoracic spinous processes are relatively narrow and have a pronounced caudal inclination. The lumbar spinous processes are rectangular and do not usually incline.

The choice of level for injection depends on the clinical problem under investigation. Where there is a localised lesion, e.g. spinal block, injection is made at the level of the lesion or into the spinous process above or below. Lower lumbar or sacral injection will fill the inferior vena cava and mid-dorsal injection will outline the azygos vein.

Procedure

A scrupulous aseptic technique should be maintained throughout the procedure. The skin is prepared in the usual way with iodine and spirit. The skin and superficial tissues over and around the chosen spinous process are infiltrated with local anaesthetic and the periosteum of the bone is also infiltrated. The skin is then nicked with a wide-bore needle or a small scalpel blade.

The bone marrow needle is introduced through the skin and soft tissues, advanced down to the bone and inclined appropriately. It is introduced through the cortex by steady pressure combined with a rotary movement. Once the needle is correctly positioned it is firmly 'gripped' by the bone and does not easily become dislodged. On withdrawing the stilet, blood should ooze from the needle; it may be necessary to aspirate a little bone marrow to achieve this and the presence of bone marrow is an important confirmatory sign of successful puncture.

The bone marrow cavity may be anaesthetised by the injection of 2–3 ml. of 2% procaine (or 1% lignocaine). (Injection is normally easy and a significant resistance indicates that the needle tip does not lie satisfactorily in the marrow.) A heparinised saline irrigation system is now attached to the needle and the solution introduced into the bone medulla, thus establishing free irrigation. The tubing should be long enough (24 inches, 60 cm.) to keep the operator's hands clear of the field of irradiation.

A test injection of 5 ml. of 45% Hypaque is made and during the injection a further lateral film with a horizontal ray is exposed. This will establish with certainty the position of the needle tip. While awaiting the film gentle intermittent irrigation is carried out. If the needle is shown to be placed satisfactorily the full injection of 20 ml. of 45% Hypaque is made and where possible 3 films are obtained at 2-second intervals. The first film is exposed towards the end of the injection and is often sufficiently informative alone. The injection should be made as rapidly as possible; some resistance is felt compared to injection for an arteriogram but it is normally possible to complete the injection within 5–7 seconds. The injection is repeated for the postero-anterior projection and it may be necessary to repeat either series or occasionally to turn the patient for a further series in the oblique position.

If injection is made into a lumbar or sacral spinous process the inferior vena cava is frequently opacified. This provides a useful approach to cavography in cases who are difficult to examine by the usual routes. It is then advisable to inject 30 ml. of contrast. When, however, the interest centres on the vertebral veins it is helpful to produce temporary occlusion of the inferior vena cava by a block or inflatable balloon placed under the abdomen; by this means contrast is diverted through the vertebral venous system.

When the examination has been completed the needle is withdrawn and a simple adhesive dressing applied. No special after-care is required and, once the patient has recovered from the premedication, he need not remain in bed.

Lower limb venography

According to the level at which it is necessary to opacify the deep veins, injection may be made into the medial malleolus at the ankle or the medial tibial or femoral condyle. For this type of injection a simple bone marrow needle is usually adequate. It should be noted that in medial malleolar puncture the cortex is entered over the upper part of the bony prominence and the needle is angled cephalad to prevent its entering the ankle joint. Before contrast is injected the position of the needle should be checked fluoroscopically. When injection of contrast has been completed 50–100 ml. of heparinised saline is infused in the attempt to 'wash out' residual contrast. At all sites injection is very painful and a general anaesthetic is required.

Pelvic venography

The increasing trend to investigation of the veins of the trunk and lower limbs is discussed on p. 264. When femoral venepuncture is impossible, venography by greater trochanteric injection under general anaesthesia is the most satisfactory alternative. The method resembles that used elsewhere but some special problems arise:—

1. The relatively slow filling of the veins tends to make the heavy contrast move into the most dependent veins. When the patient is supine this may cause failure of filling of the I.V.C. and may even give rise to appearances suggesting that there is collateral flow due to caval obstruction. It is therefore usual when examining by this method, to place the patient in the

prone position. When the examination is carried out as part of a general survey of the lower limb veins and inferior vena cava, it may be possible to enter one or other femoral vein by percutaneous puncture, with or without catheterisation. The vein which can be punctured should be catheterised first. Then if no vein can be found on the opposite side, trochanteric puncture is performed. Under these circumstances it is not ordinarily necessary to turn the patient prone, but if neither femoral vein can be punctured the patient is turned into the prone position before trochanteric puncture.

2. The trochanter may be difficult to feel and careful palpation is required. If doubt remains an opaque marker is placed against the point of proposed puncture and the area is screened to check that the localisation of the trochanter is correct.

3. The soft tissues may hinder introduction and rotation of the needle and in the 'threaded' needle designed by Hilal (M.D.) an outer cannula can be provided to allow the needle free movement as it is screwed into and through the cortex.[1]

4. The bony cortex over the trochanter is thick, necessitating a strong needle with an effective cutting tip. Suitable needles are Lea Thomas' model (M.D.) and the needle mentioned above.

5. The pressure required to traverse the cortex may move the patient sideways and an assistant on the opposite side should restrain any such movement.

6. Once the needle has been introduced and is thought to be satisfactorily positioned, this is checked by fluoroscopy. If the tip is seen to lie beneath the cortex an arterial connection is attached and a small amount of contrast is injected and watched on the screen. Provided contrast is seen to enter the venous system, injection can proceed. Should extravasation of contrast occur *the needle is left in situ* and a further puncture with a fresh needle is made. This prevents leakage of contrast out of the first puncture (Lea Thomas, 1970).

7. The injection of the two trochanters is made with *separate syringes*, (preferably pressure syringes); the use of a Y-connection results in diversion of nearly all the contrast into the side offering the least resistance (Lea Thomas, 1970). About 35–40 ml. of Conray 420 is injected on each side.

8. A film series of about 10 films at 1 per second is normally adequate. When satisfactory films have been obtained, each needle is infused with 100 ml. of heparinised saline before it is removed. A suture may be required at the site of the skin puncture.

Complications

Immediate.—If the cortex of any bone, particularly a spinous process, is punctured twice, saline or contrast invariably escapes through the first

[1] Designed by Dr. John Dow, manufactured by the London Splint Co., 50 New Cavendish Street, London W.1.

hole when injected under any pressure. A safe rule would seem to be never to penetrate a spinous process more than once at the same examination but leave an interval of at least 24 hours. In most other sites it is possible to leave a needle in the first puncture as described above, and make a second puncture with a fresh needle.

Late.—Fat emboli, bone infarcts and osteomyelitis have been recorded but are extremely uncommon and can be avoided with careful and correct technique.

References and further reading

ERHARDT, K., and KNEIP, P. (1943). *Geburtsh. u. Frauenh.*, **5**, 1.
FISCHGOLD, H., ADAM, H., ECOIFFIER, J., and PIEQUET, J. (1952). *Presse med.*, **60**, 44.
GREITZ, T., LILIEQUIST, B. and MUELLER, R. (1962). *Acta Radiol.*, **57**, 353.
ISHERWOOD, I. (1962). *Clin. Radiol.*, **13**, 73.
LEA THOMAS, M., (1970) *In Modern Trends in Diagnostic Radiology.* Edited by J. W. Maclaren p. 201, (Butterworths, London).
SCHOBINGER, R. A. (1960). *Intra-Osseous Venography* (Grune and Stratton, New York and London).

APPENDIX

Trolley setting

Basic vascular trolley setting with the following additions and alterations:—
1. An alternative form of pressure injection syringe is the T-bar model (A.H.).
2. The polyvinyl connection used should be long (24 inches, 60 cm.).
3. Bone marrow needle—'B' model spinal needle 17 S.W.G. × $1\frac{13}{16}$ inches (Shrimpton and Fletcher, Redditch). A handle for this can be made in most hospital workshops with suitable stainless steel tubing in a 'T'-shape. Lea Thomas' needle may be used for the greater trochanters.
4. Heparinised saline is 10,000 i.u. of heparin/litre.

K

CHAPTER 35

LYMPHOGRAPHY

(Peter Armstrong, M.B., F.F.R. and W. F. White, M.B., F.F.R.)

THE radiographic visualisation of lymph vessels and nodes is an investigation which can be undertaken in any X-ray Department since only a minimum of special equipment is required. Most such examinations are directed at the retroperitoneal nodes following injection of oily contrast medium into the lymphatics of the foot. The following account is based on this type of investigation, with brief references to the examination at other sites.

Applications of the method

Lymphography is used whenever it is necessary to study lymph vessels or nodes:

1. In the investigation of primary and secondary lymphoedema, chyloperitoneum and chylothorax.

2. The most important application is in the study of the distribution of malignant lymphoma. The involved nodes become enlarged, usually with a characteristic 'foamy' or 'reticular' pattern. In most cases the presence or absence of lymphomatous retroperitoneal gland involvement can be confidently diagnosed and used in the staging of these diseases when planning therapy. Follow up films make it possible to study the effects of radiotherapy and chemotherapy on the disease process. Certain inflammatory conditions, e.g. sarcoidosis, rheumatoid arthritis and ankylosing spondylitis may show a non-specific abnormal pattern which is easily confused with that due to lymphoma, but awareness of this possibility will reduce the number of false positive results obtained.

3. The method also provides the best non-operative assessment of the state of the retroperitoneal and axillary lymph-nodes in cases of known carcinoma, but it has proved too inaccurate for use in deciding the operability of the primary lesion. Post-inflammatory fibro-fatty deposits radiographically indistinguishable from small metastases have made false positive and negative diagnoses unacceptably high. It has, however, been used by some surgeons in patients requiring block dissections of regional lymph nodes, to assess the adequacy of their surgical clearance.[1] Films are

---

[1] Note: Chlorophyll can be added to the radio-opaque oil at the time of lymphography, colouring the nodes green and making them clearly visible at operation. This has the disadvantage of causing a mild inflammatory reaction which may make surgical excision more difficult. It also interferes with the passage of the radio-opaque oil through the lymphatic vessels and the subsequent pattern of opacification of the nodes, thereby reducing still further the diagnostic accuracy.

taken before closing the wound to show any nodes that may have escaped removal.

Contra-indications

1. **Iodine Sensitivity** must preclude lymphography. In cases where this is known to exist an alternative method can be utilised. In departments with access to a gamma camera, a tracer amount of radio-active gold (Au^{198}) is injected and the area concerned 'scanned'. Although this method does not give such accurate information as the routine examination, on occasions it can be extremely helpful.

2. **Infection.**—Neglected cases of lymphoedema or those with mixed venous and lymphatic oedema may have varying degrees of sepsis in the interdigital clefts. Cellulitis and/or lymphangitis are not uncommon complications of lymphoedema and should be treated before lymphography is attempted.

3. **Pulmonary Disease.**—Any diffuse pulmonary process with loss of respiratory reserve provides a source of increased risk since some degree of pulmonary oil embolisation occurs in all cases. In patients suffering from such conditions the examination should only be undertaken on the most pressing indications and care taken not to inject more contrast than is absolutely necessary.

4. **Radiotherapy to the Chest.**—A far more hazardous situation exists during or just following a course of radiotherapy to the chest, when the filtering action of the pulmonary capillaries may be lost and embolisation of the oil to the brain has been recorded. Lymphography should not be carried out during this period.

Preparation of the Patient

1. Patients with oedema of the limb are admitted into hospital some days before examination so that the oedema can be reduced by elevation and massage. Infection, particularly cellulitis and lymphangitis, may require treatment for several days and lymphography should be deferred for 10–14 days after such an attack.

Other patients need only be admitted shortly before the procedure, no special preparation being necessary.

2. The patient is examined with reference both to his general condition, particularly if an anaesthetic is required, and to the condition of the proposed site of injection. Areas in the region of the incision which are hairy will require shaving.

3. The procedure is explained to the patient who is warned about the following points:

(a) If local anaesthesia and a water-soluble medium are to be used he will experience a transient burning sensation travelling up the limb when the contrast is injected. He is warned to keep quite still and let the operator know when this sensation occurs.

(b) On return to the ward the skin over the whole body will have a bluish tinge and everything will appear as if seen through a light blue veil, due to the circulating dye. Depending on the amount of dye used, the skin regains its normal colour in one to two days but at the injection sites it will take about a week to clear completely. The urine will be coloured green for about 48 hours.

The ward staff and any visiting relatives should also be warned of these colour changes.

4. The procedure is normally performed under local anaesthesia and sedation is only required for apprehensive or restless patients. On the rare occasions when water soluble contrast is to be injected at multiple sites general anaesthesia may be preferred. A premedication will then be necessary.

5. The patient must empty the bladder before the examination.

Summary of the procedure

Pre-operative.—Treat infection and reduce oedema as much as possible.
First Stage.— 1. Inject PBV dye and massage site.
Second Stage.—2. Clean skin, towel up, inject local anaesthetic; make skin incision.
3. Expose, dissect out and cannulate lymph vessel.
4. Inject 1–3 ml. contrast medium and take film of calves to confirm adequate cannulation.
5. Continue injection until contrast reaches L4/5 disc. Remove cannula, tie lymph vessel and suture wound.
6. Obtain immediate film series and repeat in 24 hours.

Procedure

1. **Equipment.**—A lymphangiogram set is required, consisting of a fine needle (usually 27–30 S.W.G.) connected to a long flexible tube with a luer fitting on the other end. These sets may be obtained sterilised and ready for use. An injection system is also needed and must be capable of delivering between 5 and 10 ml. in an hour from two syringes. Various types of 'pump' are available. In the absence of such a pump the syringes may be held vertically in retort stands and suitable lead weights balanced on the plunger, after a preliminary trial to determine what weight should be applied to give the required flow rate. This simple apparatus, however, not uncommonly ceases to function owing to the almost inevitable variation in friction (often due to glove powder) between the syringe and the syringe barrel. Once the flow of the medium has stopped, it is not always possible to restart it. The most successful type of injector incorporates an electrically driven motor which pushes the syringe plunger forward at a constant rate. The more sophisticated injectors offer a variety of speeds.

2. **Contrast Medium.**—It is common practice to use oily contrast medium for studying both the lymph vessels and lymph nodes. The medium usually employed is Lipiodol Ultrafluid, an iodinated poppyseed oil containing 38% w/v iodine. Some workers still prefer to use water-soluble contrast media for the demonstration of the lymphatics in the limbs in the cases of lymphoedema. If this method is utilised it may be felt necessary to give a general anaesthetic as the injection of the watery media, unlike oily contrast, can be very painful. Hypaque 45 or Conray 280 are suitable water-soluble media.

3. **Coloured Dye.**—The commonly used dye is Patent Blue Violet (P.B.V.) $2\frac{1}{2}\%$ which is distributed by the manufacturers of the contrast medium. If this is not available, or in the rare cases of known sensitivity, Evans blue, indigo carmine, sky blue or aphazurine may be used instead.

4. **Assistance.**—While the experienced operator does not require an assistant it would be difficult for a beginner to accomplish a successful examination without assistance from someone skilled in the technique.

Technique

Before starting the examination the patient, having emptied his bladder, should be made as comfortable as possible and provided with suitable reading matter as he will be required to lie still for at least two to three hours.

First Stage—Injection of the dye.—The normal lymphatic is between 0·1 and 0·5 mm. in diameter, its wall and the lymph within it being colourless. In order to visualise these small vessels, it is necessary to inject a coloured dye (see above) into subcutaneous tissues. This is then carried away by the lymph which therefore becomes coloured. The skin is first cleaned with chlorhexidine/spirit and the dye is injected using a fine needle, e.g. 23 SWG, distal to the desired site of the lymphatic cannulation. In the feet the web spaces are the usual site of injection and normally about 1 ml. is used for each web space. The injection site may be gently massaged with a swab. After 10 to 20 minutes the thin blue lines denoting the course of the lymphatic vessels will usually be seen through the skin, though in fat or oedematous patients they may not be visible. The blue line is due to the dye in the lymphatic diffusing through its wall and staining the subcutaneous tissues and eventually the skin. Injection of the coloured dye into the web space between the first and second toes will demonstrate the long saphenous lymphatics: a similar injection between the fourth and fifth toes will demonstrate lymphatics on the lateral half of the dorsum of the foot, whereas an injection made just inferior to the medial malleolus will demonstrate the short saphenous vessels. In the arm, if it is required to show supratrochlear nodes, the injection should be made between the fourth and fifth fingers. If it is only necessary to demonstrate the nodes in the axilla it is usually easier to cannulate a lymphatic in the antecubital fossa, where the vessels are larger and therefore more easily found, following an injection of the coloured dye approximately 5 cm. distal to this area.

Second Stage—Cannulation of the Lymphatics.—For this stage the operator is masked and scrubs up. Gloves may be worn and the procedure is continued under aseptic conditions. The skin is again prepared with chlorexidine/spirit. The toes and the surrounding table are covered with towels.

Incision.—A suitable vessel in the dorsum of the foot is usually found near the base of the first or second metatarsal. After choosing the vessel the surrounding area is infiltrated with 1–2 ml. of local anaesthetic with or without added adrenalin* and a 1·0 cm. vertical or transverse incision is made over one of the coloured lymphatics. The choice of incision is a personal one; if a transverse incision is made and difficulty encountered in cannulating the vessel, the wound can easily be slightly extended in order to find a further vessel, whereas if a vertical incision has been used a completely new incision must be made. Having incised the skin over the lymphatic, the subcutaneous tissues are separated gently using blunt dissection, either with very fine artery forceps, or with a pair of non-pointed scissors. At this stage it is useful if an assistant is available to elevate the skin edges with skin hooks. This stretches the fine fibrous strands running in the sub-dermal region which are then easily seen and cut in the line of the main incision. The subcutaneous lymphatic vessels run immediately below these fibrous strands. Very fine lymphatics containing coloured lymph may be seen in the skin, but these should not be mistaken for the larger deeper-lying subcutaneous lymphatics—the former can be cut without flooding the incision with dye as they are small and cease oozing rapidly. The field should be kept clean and dry at all times; if there is difficulty with capillary oozing it may be helpful to lower the head of the table somewhat.

Exposure of the Vessels.—At this stage the lymphatic will be seen lying in a bed of fat and connective tissue. It is of value whilst dissecting to 'milk' lymph from the foot. This distends the lymphatic slightly and makes it more easily visible. Gentle dissection until the vessel is free of all surrounding tissue is essential to enable uniform distension of the vessel for cannulation. An unevenly distended vessel is extremely difficult to cannulate. Three pieces of 40 catgut or silk at least 6 cm. in length are passed under the vessel. The proximal and distal pieces are used to pull the distended vessel taut (Fig. 47). Milking lymph into the vessel from below with the top 'tourniquet' under slight tension will result in distension. Subsequent tension on the lower tourniquet prevents collapse of the vessel. These manoeuvres also tend to lift the vessel upwards and out of its bed enabling easier cannulation. The middle piece of catgut is loosely knotted in preparation for tying around the cannulating needle once it is in position; the

---

* Lignocaine with added adrenalin does not appear to affect the lymphatics, and has the advantage that skin suture at the end of the procedure can be accomplished without further local anaesthetic.

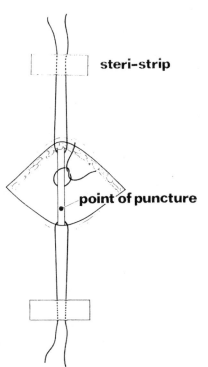

steri–strip

point of puncture

Fig. 47.—After isolating the lymphatic distension is produced by tightening the upper 'tie' and 'milking' from below. The lower tie is then tightened prior to puncture at the point indicated.

other 'ties' are not knotted since their effect is produced by simple tension.

Now a small amount of saline is injected through the lymphangiogram set to ensure that it is patent and that there are no leaks; the syringe is left connected to the tubing. The needle is then gripped between the tips of the blades of a pair of fine artery forceps near its attachment to the polythene tubing. Having distended the lymphatic with the upper catgut taut, the tension is maintained by taping this 'tourniquet' to the skin above the incision with sterile autoclave tape. Tension should then be applied and maintained in a similar fashion to the lower 'tourniquet'. The needle is aligned parallel to the lymphatic which is then punctured at the most proximal accessible point. As soon as the vessel has been punctured some of the blue lymph will pass into the polythene tubing of the lymphangiogram set. The lymphatic will naturally collapse. The needle can now be secured in place by pulling the preformed knot in the tie that is already in position (see Fig. 48); at the same time the upper tie is gently released. Before the injection of the radiographic contrast medium the cannulation must be tested to exclude leaks. If a mixture of saline and small bubbles of air is injected into the polythene tubing, the air will reveal even the smallest of leaks. Should any be discovered it will be necessary either to cannulate a fresh vessel or to attempt to pass the needle further along the lymphatic

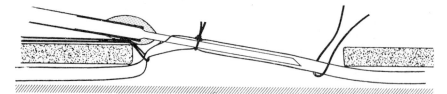

Fig. 48.—Following successful insertion of the needle the middle tie is knotted firmly and the upper tie is released.

and tie a 'tourniquet' proximal to the suspected site of leakage. It is not uncommon to have a little difficulty in the technique, particularly at first, but once the method has been mastered it becomes comparatively easy.

Injection at other Sites

It may be necessary, particularly for therapeutic lymphography, to give the injection elsewhere on the limbs. Lymphatics abound over the whole surface of both upper and lower limbs, and thus cannulation anywhere is possible, though on occasions it may be difficult. Having injected the blue dye one needs to make an incision some 5–8 cm. proximal to the injection to look for the lymphatic. In the leg and particularly in the thigh, the subcutaneous lymphatics pass almost vertically down to lie on the deep fascia. In most people this is at a considerable depth and therefore a much larger incision must be made. Whereas for conventional lymphography from the foot or wrist an incision of 1 cm only is usually required, at other sites it may be necessary to use a 5–7 cm. incision, because of the difficulties of cannulating the vessel at a depth of 3 cm. or so.

Visceral lymphography has become of interest in recent years. It is possible at operation to inject the lymphatics leaving a tumour bearing area and thereby demonstrate the lymph nodes into which they drain. For example, normally the lymph nodes draining the testicle are not visualised by lymphography from the foot, for they lie lateral to the para-aortic chain in close proximity to the renal artery. If, however, it is considered essential to demonstrate these nodes in a patient with a testicular tumour to determine whether they are involved by growth, this may be undertaken during surgery to the primary lesion. For visceral lymphography it is essential to inject the coloured material at a site which will demonstrate the lymphatics leaving the relevant organ and then to cannulate in the way that has been described above. Injections from the bowel, testicle, ovaries, uterus and bladder have been performed. Some are technically more difficult than others, but all are possible and often very informative.

Injection of the Contrast

Watery Media.—Contrast is warmed before injection. For a single leg lymphatic about 7 ml. are employed. With multiple lymphatic cannulation 3–4 ml. are injected into each cannula. Contrast is injected *as rapidly as possible*, using the smallest possible syringe. As already noted the patient is likely to experience some pain. Films are taken in quick succession, three films being required to show the whole of the leg.

If for any reason the result is unsatisfactory, further injections of contrast may be given and/or further radiographs taken, although by this time there will be some blurring due to diffusion. There may be underfilling in patients with grossly dilated and varicose lymphatics and volumes of 20 ml. or more may be required to fill the leg vessels alone.

Oily Media.—In the majority of cases an oily medium will be indicated. A slow injection at a rate approximately 5–10 ml. per hour, per lymphatic is necessary. Faster injection produces extravasation of the oil through the walls of the lymphatics.

It is essential to be sure that the vessels cannulated are indeed lymphatics and the beginner should screen or take films of the limbs after 1–3 ml. of contrast have been injected to ensure that it is not being injected into a vein. Lymphatic injection leads to the filling of multiple small vessels which, unlike veins, alter little in size as they pass up the limb. If a vein has inadvertently been injected, tiny globules of oil are seen as they travel in the blood stream. If this is detected early in the course of injection no harm will result.

The volume of Lipiodol injected is important, for any over and above that required to fill the nodes will pass into the venous system causing tiny oil emboli in the lungs. It is seldom necessary to use more than 5 ml. in each limb when injecting to show leg lymphatics and 2·5–3 ml. when injecting the arm vessels. 8 ml. of contrast in each lower limb is an average amount to show the abdominal nodes, in normal or near normal subjects. It is very useful to follow the progress of the oil by fluoroscopy or abdominal films in order to stop the injection when it reaches the level of the L4–5 disc space. Should the patient have had previous radiotherapy to the inguinal or iliac nodes, less contrast will be necessary and if the nodes are grossly enlarged they will require more oil to opacify them adequately.

Films

In cases of lymphoedema, films of the limbs are taken during the injection after 2–3 ml. of contrast have run in. In all other cases the injecting cannulae are removed and the skin sutured before the film series is exposed, as indicated below. The films are repeated the following day: whereas on the first series of films, there will be relatively little filling of the lymph nodes, most of the contrast medium being in the lymphatic vessels, on the second day good filling of the nodes with no residual contrast in the vessels will be observed. The Lipiodol will usually remain in the lymph nodes for six to twelve months, but the period can vary from one month to two years.

Films—Immediate Series

Abdomen—AP and both obliques   (in smaller patients both areas can
Pelvis—AP and both obliques   be shown on one film)
AP and lateral thoracic spine—to show the thoracic duct.

Films.—Following Day

Abdomen and Pelvis—AP, both obliques and lateral film.
Chest—for oil embolisation
Following injections into the hand or arm a similar series of films are taken showing the axillary lymph nodes.

Closure of the Incision.—At the termination of the injection the needle is removed and the lymphatic is tied off to prevent leakage of lymph. The skin is sutured after swabbing out the wound with saline. A small dressing is applied. In cases of lymphoedema the limb is enclosed in a crepe bandage from the toes to the knee (or fingers to the elbow). This is kept on for 24–48 hours.

Therapeutic Lymphography

Therapeutic lymphography has been gaining in popularity particularly in the prophylactic treatment of lymph nodes in patients with malignant melanoma and in certain cases of malignant lymphoma where other forms of treatment are either unsatisfactory or have failed. Needless to say, this should only be undertaken after consultation with the hospital physicist. This is essential since the handling of isotopes by unskilled persons may cause an x-ray room to become contaminated and unusable for a considerable length of time. The cannulation method is exactly the same but it is important to ensure that there is absolutely no leakage of the radioactive material at the injection site or elsewhere. A discussion of the indications for this treatment and the types of isotope used is beyond the scope of this book.

Complications.—In our experience, this procedure has been associated with very few complications, particularly if the contra-indications mentioned earlier are respected.

1. Occasionally there is a leakage of lymph and delay in healing. This does not occur if the lymph vessel is tied off following removal of the cannulae. It ceases with bed-rest and elevation of the limb. Escape of the contrast from lymphatics into the calf sometimes occurs with too rapid injection. This does not appear to cause any harm.

2. Lymphangitis and/or cellulitis have been reported. They should not occur with adequate technique. Treatment is with bed-rest, limb elevation and antibiotics.

3. Sensitivity reactions caused by the contrast, the coloured dye and the local anaesthetic, may sometimes occur. They may either be a local erythema and irritation or a general reaction requiring antihistamines, adrenaline or steroids.

4. Minor degrees of pulmonary oil embolism are common but seldom produce symptoms. More severe embolisation can occur with accidental intravenous injection or with the use of excessive volumes of contrast. As already noted, patients with diffuse disease or large deposits are particularly at risk.

5. On rare occasions the patient develops breathlessness apparently due to pulmonary oedema a short while after the end of the injection. It is not related to the volume of oil injected and may be due to some form of 'hypersensitivity'; it seldom lasts more than 2–3 hours.

6. An unexplained fever without other signs is occasionally encountered.

7. Delayed complications such as lymphoedema have been reported after oily contrast media.

Lymphogram Setting

Part I

Injection of the staining dye between the toes:

5 ml. syringe
Nos. 20 and 17 intramuscular needles
Drawing up cannula
Sterile gauze squares 4″ × 4″
Chlorhexidine/spirit
Ampoules of Patent Blue V
Inj. Lignocaine 1%

Part II

Exposing the lymph vessels, injecting the contrast and suturing the skin incisions.

Sterile trolley.

Three dressing towels. Guaze and wool swabs. Two gallipots.
Receiver.
Sponge holding forceps
Towel clips
Bard parker handle no. 3
No. 15 blade
McIndoes dissecting forceps
Ophthalmic dissecting forceps, non-toothed
Ophthalmic scissors, straight with pointed blades
Opthalmic scissors, straight with round blades
4 pairs Mosquito artery forceps, curved.
2 pairs Mosquito artery forceps, straight
2 Aneurysm hooks
Needle Holder
20 ml. syringe
10 ml. syringe
Nos. 1 and 17 intramuscular needles } for local anaesthetic
Drawing up cannula
2 × 10 ml. non-lubricated syringes, Luer lock
Drawing up cannula
2 × disposable lymphangiogram sets (Macarthy's Ltd., Romford, Essex)
Sterile autoclave tape
Sterile emery paper, fine Kifa size 4/0*
Lengths of 4/0 black silk or catgut
Skin suture needle and silk

* Some workers like to reduce the sharpness of the needle with fine emery paper, believing that it prevents accidental penetration of the lymphatic after the needle has been inserted.

Lower shelf

Normal saline
Lipiodol Ultrafluid ampoules
Lignocaine 1% (with or without added adrenalin)
Good spot light
Magnifying lamp
Pump for injection of contrast, 10 ml. in 1½ hours

Occupational therapy for patient, e.g. reading, conversation, radio

Suggested Reading

BURN, J. I., and BOHRER, S. P. (1965). *Br. J. Cancer*, **19,** 321–329.
FISCH, U. P. (1964). *Jnl. of Laryngol. and Otol.*, **78,** 715–726.
FRAIMOW, W., WALLACE, S., LEWIS, P., GREENING, R. R., and CATHCART, R. T. (1965). *Radiology*, **85,** 231–241.
HARTGILL, J. C. (1964). *Jnl. of Obstet. and Gynae. of Brit. Comn.*, **71,** 835–853.
HRESHCHYSHYN, M. N., SHEEHAN, F. R., and HOLLAND, J. F. (1961). *Cancer*, **14,** 205–209.
KENDALL, B. (1964). *Radiography*, **30, 351,** 79–87.
KINMONTH, J. B. (1952). *Clin. Sci.*, **2,** 13–20.
KINMONTH, J. B. (1954). *Ann. Roy. Coll. Surgeons (Eng.)*, **15,** 300–315.
MACDONALD, J. S., and WALLACE, F. C. W. K. (1965). *Br. J. Radiol.*, **38,** 93–99.
PRATT, P. P., and ABBES, M. (1964). *Cancer*, **17,** 850–855.
VIAMONTE, M. (1964). *Acta Radiologica*, **2,** 394–400.

GENERAL APPENDIX

Needle sizes.—The methods used for designating needle size are liable to cause confusion, principally because the two systems in common use both employ numbers. These numbers are quite unrelated so that a hypodermic No. 20 needle is very different from a 20 gauge needle. The 'gauge' system (more fully Standard Wire Gauge or S.W.G.) indicates the external diameter of a needle and *any* needle can therefore be referred to in terms of its length and gauge number. In practice this is usually only used for the larger needles and most needles of 2 inches length or less belong either to the 'hypodermic' or the 'serum' range of numbers. These are both series in which arbitrary numbers are allotted to needles of certain fixed diameters and lengths. Thus a hypodermic No. 20 needle is an intradermal needle, 26 S.W.G. × $\frac{5}{8}$ in. and a hypodermic No. 1 is the commonest intramuscular needle; its dimensions are 21 S.W.G. × $1\frac{1}{2}$ ins. In Tables I, II and III the full range of measurements is set out. In many cases the difference in size between needles is insignificant and this has been recognised in the new British Standard (3522:1962) for hypodermic needles. Table IV is reproduced from B.S. 3522:1962 and shows the suggested size range of hypodermic needles. It will be noted that this achieves a considerable reduction in the number of needles and that it defines needle size in terms of either gauge and length in inches or the approximate metric dimensions. The widespread use of disposable needles has led to a welcome diminution in the variety of needle sizes, though the sizes offered do not always correspond to the proposed British Standard range shown in Table V.

There are a number of terms used in describing needles which merit further discussion:—

Hypodermic needles.—Strictly this should mean a needle for subcutaneous injection but the word 'hypodermic' has acquired at least two further meanings. It is used to distinguish needles for injection from suture needles, as in the phrase 'Hypodermic Surgical Mounted Needles'. It is also given, as mentioned above, to the hypodermic size numbers for needles, as in 'hypodermic No. 1'.

Serum needles.—The word 'serum' in connection with needles also has two meanings. There are the 'serum sizes' numbers 0–VI shown in Table III. The word 'serum' may sometimes be applied to a needle of less than 3 inches length when the needle does not belong to any of the hypodermic sizes.

Exploring needles.—By common practice these are needles of 3 inches or more in length.

TABLE I
STANDARD WIRE GAUGE MEASUREMENTS

S.W.G.	Ext. diam. mm.	Ext. diam. ins.	Corresponding needle type
12	2·642	0·104	Robb-Steinberg needle
13	2·337	0·092	Lea Thomas' intraosseous needle
14	2·032	0·080	Seldinger needle: size PE 205
15	1·829	0·072	} Range of aortogram needles
16	1·626	0·064	
17	1·422	0·056	Seldinger needle: size PE 160
18	1·219	0·048	Sheldon arterial cannula (thin wall)
19	1·016	0·040	Thin wall needle for injecting large volumes of contrast (p. 47)
20	0·914	0·036	
21	0·813	0·032	Intramuscular needles
22	0·711	0·028	
23	0·610	0·024	
24	0·559	0·022	Subcutaneous and intradermal needles
25	0·508	0·020	
26	0·457	0·018	
27	0·416	0·0164	Needles for lymphography
28	0·376	0·0148	

Note.—1. The gauges employed in U.S.A. Federal Specifications correspond closely with the dimensions given above.

2. There are other wire gauges in use (e.g. Birmingham Wire Gauge) with slightly differing dimensions. The Standard Wire Gauge is the only legal wire gauge in the United Kingdom.

TABLE II
HYPODERMIC NEEDLE SIZES

Hypodermic size number	S.W.G.	Length inches	Metric Dimensions mm.
0	20	$1\frac{5}{8}$	0·90 × 41·5
1	21	$1\frac{1}{2}$	0·80 × 38
2	22	$1\frac{5}{16}$	0·70 × 33
12	23	$1\frac{3}{16}$	0·65 × 30
14	23	$1\frac{3}{16}$	0·60 × 30
15	23	1	0·60 × 25
16	24	1	0·55 × 25
17	25	$\frac{15}{16}$	0·50 × 23·5
18	26	$\frac{3}{4}$	0·45 × 19
19	26	$\frac{11}{16}$	0·45 × 17·5
20	26	$\frac{5}{8}$	0·45 × 15·5

TABLE III
SERUM NEEDLE SIZES

Serum size number	S.W.G.	Length inches	Metric Dimensions mm.
0	17	$2\frac{3}{8}$	1·45 × 60
I	18	2	1·25 × 50·5
II	19	2	1·10 × 50·5
III	20	2	0·90 × 50·5
IV	21	2	0·80 × 50·5
V	22	2	0·70 × 50·5
VI	23	2	0·65 × 50·5

SPECIFICATIONS FOR HYPODERMIC NEEDLE SIZES FROM B.S. 3522:1962[1]

Designated size nom. dia. × nom. length (in.) (Gauge)	Designated size nom. dia. × nom. length (mm./10) (mm.)	External diameter of needle tube				Minimum bore of needle tube		Length of needle	For information only Comparison with British Hypodermic and Serum ranges of sizes
		Max. in.	Min. in.	Max. mm.	Min. mm.	in.	mm.	mm.	
26G × ½	5 × 13	0·018 5	0·017 5	0·470	0·445	0·009 5	0·241	12·5±1·0	Hypo. No. 20
26G × ¾	5 × 20	0·018 5	0·017 5	0·470	0·445	0·009 5	0·241	20·0±1·0	Hypo. No. 18
26G × 1	5 × 25	0·018 5	0·017 5	0·470	0·445	0·009 5	0·241	25·0±1·0	Hypo. No. 17
23G × 1	6 × 25	0·025 5	0·024 5	0·648	0·622	0·012 5	0·318	25·0±1·0	Hypo. No. 15
23G × 1¼	6 × 32	0·025 5	0·024 5	0·648	0·622	0·012 5	0·318	31·5±1·5	Hypo. No. 12 and 14
23G × 2	6 × 50	0·025 5	0·024 5	0·648	0·622	0·012 5	0·318	50·0±1·5	Serum No. 6
21G × 1½	8 × 40	0·032 5	0·031 5	0·826	0·800	0·019 5	0·495	40·0±1·5	Hypo. No. 1
21G × 2	8 × 50	0·032 5	0·031 5	0·826	0·800	0·019 5	0·495	50·0±1·5	Serum No. 4
19G × 2	10 × 50	0·043 5	0·040 5	1·105	1·029	0·025 5	0·648	50·0±1·5	Serum No. 2
18G × 2	12 × 50	0·050 5	0·047 5	1·283	1·207	0·031 5	0·800	50·0±1·5	Serum No. 1

Note.—The diameters and lengths in millimetres of the needles given in this table are based on the R/10 series of preferred numbers based on a log-arithmic series in accordance with B.S. 2045, 'Preferred numbers'.

[1] Extracts from B.S. 3522: 1962 are reproduced by permission of the British Standards Institution, 2 Park Street, London, W.1., from whom copies of the complete standard may be purchased.

TABLE V

SIZE RANGE OF HYPODERMIC NEEDLES FOR SINGLE USE

Designated size (metric units) nom. dia. × nom. length (mm./10) × (mm.)	Nearest equivalent size (inch units) nom. dia. × nom. length (gauge) × (in.)	External diameter of needle tube				Minimum bore of needle tube		Length of needle	Comparison with British Hypodermic and Serum ranges of sizes
		Max. mm.	Min. mm.	Max. in.	Min. in.	mm.	in.	mm.	
5 × 16	25G × ⅝	0·51	0·495	0·020 5	0·019 5	0·241	0·009 5	16±1·0	Hypo. No. 18
6 × 25	23G × 1	0·648	0·622	0·025 5	0·024 5	0·318	0·012 5	25±1·5	Hypo. No. 15
8 × 40	21G × 1½	0·826	0·800	0·032 5	0·031 5	0·495	0·019 5	40±2·0	Hypo. No. 1
10 × 40	19G × 1½	1·105	1·029	0·043 5	0·040 5	0·648	0·025 5	40±2·0	—

TABLE VI
POLYTHENE TUBING
(Selected sizes)

Intra-medic Catalogue No.	Portex[1] Catalogue Ref. No.	Diameter in mm.		Area of cross-section of lumen in sq. mm.	Diameter in inches		Needle gauge which fits into tubing
		Internal	External		Internal	External	
PE 160	(O)PP 160	1·14	1·57	1·021	0·045	0·062	18
PE 190	PP 190	1·19	1·70	1·127	0·047	0·067	18
—	PP 120	1·00	2·00	0·785	0·039	0·078	19
PE 200	PP 200	1·40	1·90	1·539	0·055	0·075	17
PE 205	(O)PP 205	1·57	2·08	1·950	0·062	0·082	16
—	PP 204	1·55	1·90	1·882	0·061	0·075	16
—	(O)PP 202	1·50	2·70	1·767	0·059	0·106	16
PE 240	(O)PP 240	1·67	2·42	2·180	0·066	0·095	15
PE 260	PP 260	1·77	2·80	2·477	0·070	0·110	15
KIFA							
Red		1·2	2·2	1·13	—	—	—
Green		1·2	2·4	1·13	—	—	—
Yellow		1·5	2·8	1·767	—	—	—
Grey		1·8	2·8	2·545	—	—	—

[1] Catheters with the prefix 'O' are also available in opaque polythene. Since going to press the catalogue numbers of the Portex range have been altered, and are now as follows

PP 160—800/100/320
PP 190—800/100/340
PP 120—800/100/300
PP 200—800/100/360
PP 205—800/100/420
PP 204—800/100/400
PP 202—800/100/380
PP 240—800/100/460
PP 260—800/100/480

OPP 160—800/105/320
OPP 205—800/105/420
OPP 202—800/105/380
OPP 240—800/105/460

TABLE VII
RANGE OF LONGDWEL TEFLON CATHETER-NEEDLES (B.D.)

Catalog Number	Description	Thin-wall Catheter	Catheter I.D.	Inner Needle Gauge	Bevel	Inner Needle Stylet	Obturator for Catheter
6710	01-0043	20G 2½″	·025″	23G	Short	None	None
6717	01-0043	18G 2½″	·042″	19G	Short	None	None
6716	01-0044	18G 2½″	·042″	19G	Short	None	Teflon
6715	01-0045	18G 2½″	·042″	19G	Arterial	Fitted to Bevel	Teflon
6726	01-0046	16G 2½″	·050″	18G	Arterial	Fitted to Bevel	Teflon
6731	01-0047	15G 2½″	·059″	17G	Arterial	Fitted to Bevel	Teflon
6711	01-0048	20G 4″	·025″	23G	Spinal	Fitted to Bevel	Teflon
6719	01-0048	18G 4″	·042″	19G	Spinal	Fitted to Bevel	Teflon
6732	01-0049	15G 4″	·059″	17G	Short	None	None
6721	01-0050	18G 6″	·042″	19G	Arterial	Fitted to Bevel	Teflon
6723	01-0050	18G 8″	·042″	19G	Arterial	Fitted to Bevel	Teflon
6712	01-0051	20G 6″	·025″	23G	Arterial	Fitted to Bevel	Metal
6713	01-0051	20G 8″	·025″	23G	Arterial	Fitted to Bevel	Metal
6733	—	16G 4″	·050″	18G	Arterial	Fitted to Bevel	Metal
6734	—	16G 6″	·050″	18G	Arterial	Fitted to Bevel	Metal
6735	—	16G 8″	·050″	18G	Arterial	Fitted to Bevel	Metal

TABLE VIII

IODINE CONTENT OF WATER-SOLUBLE CONTRAST MEDIA

Contrast medium	% concentration	Iodine Content G./ml.	Viscosity cps. at 37°C	Chemical composition
Hypaque	25	0·15	2·1	Sodium diatrizoate: the sodium salt of DIATRIZOIC ACID (N,N′diacetyl-3, 5-diamino-2:4:6 triiodobenzoic acid).
	45	0·27	8·3	
	65	0·39	12·2	25% w/v. of the sodium salt and 50% of the N-methylglucamine salt of diatrizoic acid.
	85	0·44		28·33% w/v. of the sodium salt and 56·67% of the N-methylglucamine salt of diatrizoic acid.
Urografin	30	0·15	1·4	Each contains a mixture of the sodium and methylglucamine salts of diatrizoic acid in the proportion of 10 to 66.
	45	0·22		
	60	0·29	4·0	
	76	0·37	8·5	
Angiografin	65	30·58	5·1	Methylglucamine diatrizoate
Urovison	58	32·5	3·5	A mixture of the sodium and methylglucamine salts of diatrizoic acid in the proportion 40:18
Retro Conray	35	0·163		Methylglucamine iothalamate
Conray 280	60	0·28	4·0	Sodium iothalamate ,, ,,
325	54	0·325	2·75	
Cardio-Conray	—	0·4	8·6	Methylglucamine iothalamate 52% w/v. and sodium iothalamate 26% w/v.
Conray 420	70	0·42	5·4	Sodium iothalamate ,, ,,
480	80	0·48	8·0	
Triosil	25	0·146	—	Sodium metrizoate: the sodium salt of METRIZOIC ACID (3-acetamido-2:4:6-tri-iodo-5-N-methylacetamidobenzoic acid), with traces of the calcium and magnesium salts.
	45	0·26	2·0	
	60	0·35	—	
	75	0·44	6·4	

TABLE IX
IODINE CONTENT OF OTHER CONTRAST MEDIA

Umbradil viscous U.	35	0·175	Diodone 35% in carboxymethyl cellulose jelly.
Biligrafin Biligrafin forte	30 50	0·15 0·25	Methyl-glucamine salt of N,N′-adipic-di-(3-amino-2:4:6—triiodobenzoic acid).
Endografin	70	0·35	,,
Endografin FL	50	0·25	,,
Dionosil Oily	60	·34 (approx.)	Propyliodone (the n-propyl ester of 3:5-diiodo-4-pyridone-N-acetic acid) suspended in arachis oil.
Dionosil Aqueous	50	·3 (approx.)	Propyliodone in aqueous suspension.
Hytrast	—	·5	Iopydol/Iopydone in sodium carboxthymethyl cellulose 1·5% suspension in water
Lipiodol viscous	—	·4 (approx.)	Iodised ethyl esters of fatty acids of poppyseed oil.
Lipiodol ultra-fluid	—	·38 (approx.)	,, ,, ,,

TABLE X
LIGNOCAINE—RECOMMENDED MAXIMUM DOSES
ml. (cc.) lignocaine solution

Body weight	Kg. Lb.	10 22	20 44	30 66	40 88	50 110	60 132	70 154	80 176
Lignocaine (plain)	0·5% 1·0% 1·5% 2·0% 4·0%	6	11 6	17 9 5	23 11 7 5	29 14 9 7 3	34 17 11 9 4	40 20 13 10 5	46 23 15 11 6
Lignocaine (with adrenaline)	0·5% 1·0% 2·0%	14 7 4	29 14 7	43 21 11	57 29 14	71 36 18	86 43 21	100 50 25	115 57 29

(By permission of Duncan, Flockhart and Co. Ltd., Edinburgh.)

TABLE XI
PRILOCAINE (CITANEST)—RECOMMENDED MAXIMUM DOSES
Figures given are for a healthy adult of about 70 kg. body weight. As with lignocaine, proportionately less is given to smaller subjects and to debilitated and aged patients.

ml. prilocaine solution

%	Plain	With adrenaline
0·5	80	120
1·0	40	60
1·5	26·6	40
2·0	20·0	30
3·0	13·3	20
4·0	10·0	15·0
5·0	8·0	12·0
10·0	4·0	6·0

INDEX